Timna-Josua Kühn

Characterization of Metallic Nanoparticles by X-ray Spectroscopy

Timna-Josua Kühn

Characterization of Metallic Nanoparticles by X-ray Spectroscopy

Introduction and Application of High-Resolution X-ray Absorption and Emission Spectrocsopy

Südwestdeutscher Verlag für Hochschulschriften

Impressum/Imprint (nur für Deutschland/only for Germany)
Bibliografische Information der Deutschen Nationalbibliothek: Die Deutsche Nationalbibliothek verzeichnet diese Publikation in der Deutschen Nationalbibliografie; detaillierte bibliografische Daten sind im Internet über http://dnb.d-nb.de abrufbar.
Alle in diesem Buch genannten Marken und Produktnamen unterliegen warenzeichen-, marken- oder patentrechtlichem Schutz bzw. sind Warenzeichen oder eingetragene Warenzeichen der jeweiligen Inhaber. Die Wiedergabe von Marken, Produktnamen, Gebrauchsnamen, Handelsnamen, Warenbezeichnungen u.s.w. in diesem Werk berechtigt auch ohne besondere Kennzeichnung nicht zu der Annahme, dass solche Namen im Sinne der Warenzeichen- und Markenschutzgesetzgebung als frei zu betrachten wären und daher von jedermann benutzt werden dürften.

Coverbild: www.ingimage.com

Verlag: Südwestdeutscher Verlag für Hochschulschriften GmbH & Co. KG
Heinrich-Böcking-Str. 6-8, 66121 Saarbrücken, Deutschland
Telefon +49 681 37 20 271-1, Telefax +49 681 37 20 271-0
Email: info@svh-verlag.de

Approved by: Bonn, Universität Bonn, Dissertation, 2011

Herstellung in Deutschland (siehe letzte Seite)
ISBN: 978-3-8381-3350-8

Imprint (only for USA, GB)
Bibliographic information published by the Deutsche Nationalbibliothek: The Deutsche Nationalbibliothek lists this publication in the Deutsche Nationalbibliografie; detailed bibliographic data are available in the Internet at http://dnb.d-nb.de.
Any brand names and product names mentioned in this book are subject to trademark, brand or patent protection and are trademarks or registered trademarks of their respective holders. The use of brand names, product names, common names, trade names, product descriptions etc. even without a particular marking in this works is in no way to be construed to mean that such names may be regarded as unrestricted in respect of trademark and brand protection legislation and could thus be used by anyone.

Cover image: www.ingimage.com

Publisher: Südwestdeutscher Verlag für Hochschulschriften GmbH & Co. KG
Heinrich-Böcking-Str. 6-8, 66121 Saarbrücken, Germany
Phone +49 681 37 20 271-1, Fax +49 681 37 20 271-0
Email: info@svh-verlag.de

Printed in the U.S.A.
Printed in the U.K. by (see last page)
ISBN: 978-3-8381-3350-8

Copyright © 2012 by the author and Südwestdeutscher Verlag für Hochschulschriften GmbH & Co. KG and licensors
All rights reserved. Saarbrücken 2012

Acknowledgements

I would like to express my sincere gratitude to the following persons without whom this work had not been possible:

Prof. Dr. J. Hormes for providing me the opportunity to work at the Institute of Physics at the University of Bonn, as well as at the Canadian Light Source, and for being able to write my PhD thesis with a lot of freedom at the same time, on this most fascinating topic.

Prof. Dr. S. Linden for undertaking the task of co-supervisor as well as Prof. Dr. M. Drees and Prof. Dr. J. Beck for accepting to be the two remaining members of the Examining Board.

Prof. Dr. H. Bönnemann and Dr. N. Matoussevitch of STREM Chemicals GmbH, as well as Dr. G. Khelashvili of the FZK, for providing the metallic nanoparticles, essential for my work.

Dr. H. Schulenburg for sharing his knowledge and results about the cobalt-platinum catalyst.

Dr. J. Rothe and Dr. B. Brendebach for introducing me to the instrumental part of X-ray absorption spectroscopy in practice and for their great support at the INE beamline at the FZK, sometimes even at unusual times.

Dr. W. Caliebe for his superior support at the W1 beamline at HASYLAB and his numerous advices regarding experiment as well as theory.

Dr. P. Glatzel for his outstanding helpful discussions regarding site-selective X-ray absorption spectroscopy. Further on, for sharing his personal beamtime at the ESRF with me, thus allowing me to perform my most important measurements.

The whole SYLI group, that is those that already left – Dr. T. Vitova, Dr. S. Zinoveva, Dr. H. Lichtenberg and all the others, for introducing me to the exciting field of X-ray absorption spectroscopy – as well as the few still present: G. Bovenkamp and M. Niestroj for friendship and fruitful discussions and, in particular, for the conduction of measurements in my stead at the time when my son was born. I also thank M. Niestroj for having been my guide in the distant city of Saskatoon during my visits at the Canadian Light Source.

All my friends and my family for help, support and encouragement during this long period, and special thanks goes to my parents for always seeing more in me than I do.

Finally, I thank my wife and my son for never stop believing in me and for being the life in my life.

Contents

1. **Introduction**   1
2. **Theory of X-ray Absorption and X-ray Emission Spectroscopy**   5
   2.1. Interactions of photons with matter 5
   2.2. X-ray absorption spectroscopy (XAS) 8
      2.2.1. X-ray absorption fine structure (XAFS) 9
      2.2.2. Multiple scattering formalism 11
      2.2.3. The EXAFS equation 16
   2.3. X-ray emission spectroscopy (XES) 17
      2.3.1. High-resolution fluorescence detected XAS (HRFD-XAS) 19
      2.3.2. Selectivity of HRFD-XAS 22
      2.3.3. Lifetime influence onto HRFD-XAS 23

3. **Experiment**   25
   3.1. The synchrotron radiation facility 25
   3.2. ANKA and the standard XAS experiment 27
   3.3. HASYLAB and the RIXS experiment 29
   3.4. ESRF: RXES and HRFD-XAS experiments 31

4. **Synthesis of Nanoparticles**   33
   4.1. Cobalt nanoparticles 34
      4.1.1. Co nanoparticles for high-resolution X-ray measurements 35
   4.2. Cobalt-Platinum nanoparticles 36
      4.2.1. Co_3Pt/C nanoparticles as catalysts 36

5. **Co-Pt Nanoparticles as Catalysts in Fuel Cells**   39
   5.1. XANES measurements of Co_3Pt/C nanoparticles 41
   5.2. EXAFS measurements of Co_3Pt/C nanoparticles 44
   5.3. Conclusion 50

6. **Site-Selective XAS**   55
   6.1. General strategy for site-selective XAS 56
   6.2. 1s3p-RIXS at wiggler beamline W1 56
      6.2.1. Co-CoO test system 57
      6.2.2. Co nanoparticles 65
   6.3. 1s3p-RIXS, HRFD-XANES and VTC-XES at undulator beamline ID26 . 72
      6.3.1. Experimental 72
      6.3.2. Overview 72
      6.3.3. NRXES $K\beta_{1,3}$ spectra 74
      6.3.4. Valency/Site-selective XANES 76

	6.3.5. Valency/Site-selective EXAFS	91
	6.3.6. Valence-to-core spectroscopy (K$\beta_{2,5}$)	102
	6.3.7. HRFD-XANES from K$\beta_{2,5}$	108
	6.3.8. Summary of Co nanoparticle properties	112
6.4.	Conclusion	112

7. Summary and Outlook — 115

Appendices — 117

A. Simple Models — 119

B. Singular Value Decomposition — 123

C. XANES — 125

D. EXAFS — 131

Bibliography — 149

1. Introduction

Nowadays, the study of nanoparticles – nanotechnology – is a hot topic as respective applications already have a crucial part in almost all areas of technology, such as medicine, electronics (especially computers), energy, environment, fabrics, food and so forth [93, 102]. Here, metallic nanoparticles are of particular interest due to their special thermal, electronic, magnetic and optical properties. Iron or gold-iron nanoparticles, for example, are directly injected into tumours which they destroy by being heated up, with the help of a high-frequency magnetic field [90] or a near-infrared laser [44]. But what are nanoparticles and what makes them so special? The key to answer that question is found in their size: nanoparticles are very small from the viewpoint of solid state physics but (in general) big with respect to atomic/molecular physics. As soon as at least one dimension of a particle is between 1 and 100 nm, it is called a nanoparticle [103] and two effects become important: (1) Quantum effects due to the confinement of the electrons [15] and (2) surface effects as the ratio of surface atoms to bulk atoms increases drastically. For spherical cobalt nanoparticles of diameters 100, 10, 4, and 2 nm, for example, this ratio can be estimated to 1%, 10%, 25% and 50%.[1] As a result, most of the nanoparticles' properties become size-dependent and even fascinating new properties arise: "Opaque substances become transparent (copper); stable materials turn combustible (aluminium); insoluble materials become soluble (gold) and a material such as gold, which is chemically inert at normal scales, can serve as a potent chemical catalyst." [104] However, sometimes even undesirable effects occur, some ferroelectric materials, for example, lose their ferroelectric properties when they become nanosized [66].

It is of paramount importance thus, to properly control and understand the nanoparticles' special properties. A lot of chemical synthesis routes already exist to control the production of, e.g., metallic nanoparticles of any desired size and shape. Various stabilizing agents and surfactants are used for that reason, and it is beyond dispute that these agents do strongly determine not only the size and shape, but also the geometrical, magnetic and electronic properties of the final product. It has become evident here that there is a strong interdependence of all the nanoparticles' properties and that it is quite difficult to change one property in a systematic manner without influencing the others [37, 114]. The surfactants, moreover, could also serve as final protective coating, as metal nanoparticles are highly reactive due to their small size, and must be preserved from unrestricted agglomeration and complete oxidation when exposed to air in real life applications. Alternatively, the nanoparticles are put onto a substrate or a coating is attached after the synthesis whereby its functional groups could interact weakly or strongly with the nanoparticles, thereby preserving the electronic properties of the "naked" nanoparticles or altering them by charge redistribution, so that new (wanted) properties can arise, like self-assembling [92] (compare [114]). In the past years, strong indications were gathered that these coatings (or surface layers) are not just influencing but rather determining the nanoparticles' final properties [37, 114]. Consequently, a strong demand exists for being able to understand the intrinsic interplay of the nanoparticles' coating and its interior, which necessitates respective experimental techniques, capable of distinguishing between both parts.

Among the most common techniques for material science are (high-resolution) transmission elec-

---

[1] Simple geometry, see appendix A.

1. Introduction

tron microscopy (HR)TEM for imaging and size determination [109], and scanning TEM (STEM) combined with energy-dispersive X-ray spectroscopy (EDX) for obtaining the elemental composition [65]. All these are valuable, if not mandatory, information that, however, do not distinguish between bulk and surface of a particle. In order to get access to crystallographic properties of materials, X-ray diffraction (XRD) is the tool of choice since 1913, when W. H. Bragg and W. L. Bragg discovered that crystals produce an ordered diffraction pattern when irradiated with X-rays [14]. Such a pattern shows characteristic peaks in dependence on the lattice phase and its spacings. However, XRD has a drawback when it comes to really small particles due to its dependence on periodic long-range order. The consequence is an increase of the peaks' widths with decreasing particle size, making them unspecific at a certain point – somewhere at about a few nanometers of particle size, albeit it is very challenging already below 10 nanometers. Further on, standard XRD is insensitive to surfaces, which can be circumvented, however, by so-called grazing incidence X-ray diffraction (GIXRD) measurements, where the crystal structure of the surface is probed by decreasing the angle of incidence [9]. Thus, a combination of XRD and GIXD can be an important step to understand something about the interplay of coating and bulk of a nanoparticle, though only for the crystalline parts. To describe para- or even non-crystalline materials as well, the angle of incidence has to be decreased even more as this is accompanied by a decrease of the sensitivity scale, from atomic resolution (XRD) towards the nanometer regime, and the dependence on mere medium-range order. Consequently, this type of experiments – small angle X-ray scattering SAXS [24] – yields only the morphology of the particles and when applied with grazing incident angles (GISAXS) [57], the morphology of surface structures. Alternatively to X-rays, electrons can be utilized for spectroscopy, too. In particular to get crystal properties of a particles' surface, diffraction based techniques with electrons, which are less deeply penetrating the material under investigation (compared to X-rays) as they do interact more strongly with it, are possible. Two methods, either the use low-energy electrons or of high-energy electrons with a very small angle of incidence, are feasible to probe the top-most layers only. Both methods, low-energy electron diffraction (LEED) [64] and reflection high-energy electron diffraction (RHEED) [39], however, require very "clean" crystalline surfaces. The cleaning can be realized chemically by etching or physically by cleavage, or alternatively some passive coating can be attached, which is all impossible for nanoparticles as its alters their properties. Lastly, X-ray photoelectron spectroscopy (XPS), which also is surface sensitive, allows to determine the elemental composition and the chemical as well as electronic state of each element and does not presume anything about the conditions of the surface, besides it is element-specific. Here the "surface" is in general the top 1 to 10 nm of the material, i.e. for nanoparticles possibly the whole particle. However, XPS and the aforementioned electron spectroscopies likewise, require (ultra) high-vacuum, as otherwise photoelectrons get lost due to their interaction with molecules of the air which, furthermore, can get adsorbed to the materials surface, thereby modifying its properties.

Being independent on "special conditions" and being likewise sensitive to crystalline and amorphous materials of any size, X-ray absorption spectroscopy (XAS) is a preferential tool for the nanoregime. In a XAS experiment, X-rays emerging from a synchrotron, are tuned through an "absorption edge" of the element of interest, and the absorption is detected in dependence on the X-ray energy. These absorption edges are due to the ejection of a deep-bound electron and are characteristic for each element, making XAS element-specific, which is important when dealing with, for example, bi- (or multi-) metallic systems. For transition metals like iron, cobalt, or the heavier platinum and gold, hard X-rays are needed that exhibit high penetration depths, so that no vacuum is necessary, making XAS a technique of choice for in-situ experiments, for example, to keep track of a complete synthesis procedure [54]. XAS is divided into the near-edge X-ray absorption fine structure (or X-ray absorption near edge structure, XANES) that provides (mainly) information about the electronic and geometric structure of the chosen element and the extended X-ray absorption fine structure (EXAFS) by which the local atomic environment from about 0 to 10 Å around the chosen atoms is accessible. However, despite these advantages of XAS, it can in general not function as the sole technique to completely characterize a nanoscaled system but rather as a valuable complement to other techniques, since, for example, it barely gives information about particle sizes and does not allow to distinguish between light elements of the same period that are attached to the metal.

Both XANES and EXAFS provide element-specific insight into materials, however, by summing up the absorption from all different trapping sites or chemical forms of the chosen element. Thus, a XAS spectrum is a superposition of XAS spectra from the different sites and if these are precisely known and also available for measurement, it is possible to describe them separately. Depending on the aim of an experiment it also could be that a separation is not necessary, for example, when the characterization of the material in its entirety, like it will be utilized in the final application, is needed. However, more often the question arises why some characteristic properties are present and for the case that the different chemical forms are a priori unknown – which is often the case in particular for nanoparticles – XAS will not allow to answer that question. A solution is given here by the pioneering work of M. M. Grush in 1995 *et al.* [30] by combining XAS and XES (X-ray emission spectroscopy). They detected X-ray fluorescence which is emitted subsequent to an X-ray absorption event and therefore is proportional to it, from the respective emission peaks of elements in different trapping sites. As these fluorescence peaks show a shift, dependent on the chemical form, the XAS spectra obtained in this manner are (partially) selective to these forms of the element. M. M. Grush *et al.* succeeded in recording (partially) site-selective EXAFS, though they still suffered at that time from low signal-to-noise ratio. This kind of work was successfully continued in the site-selective EXAFS measurements on the first multivalent compound ever synthesized, the well-known Prussian Blue, by P. Glatzel *et al.* in 2002 [26], with the help of improved technical devices as well as a 3rd-generation synchrotron source.

In this thesis, XAS is chosen to characterize metallic nanoparticles, as it provides insight into the electronic structure via XANES and into the local atomic environment via EXAFS, independently on size, shape and state of crystallinity of the surface and the interior of a nanoparticle. Besides, if suitable references and/or model simulations are available, the specific compound or phase (if crystalline) is identifiable. Moreover, the extension of XAS (via XES), site-selective XAS, enables to get the aforementioned kind of information for the metal connected to the coating and the metal in the bulk separately, which in principle allows to identify the interplay of these two sites, thereby giving pathways at hand to further improve and tailor the nanoparticles' properties.

The structure of this thesis is as follows: Subsequent to this introduction the theory of XAS will be derived in chapter 2, and a brief introduction to the synchrotron, as well as a description of the experimental stations hosted there to realize XAS experiments will be given in chapter 3. This is followed in chapter 4 by a concise treatment of nanoparticle synthesis in general, as well as for the specific nanoparticles investigated in this work. In chapter 5 eventually, a nanoparticle catalyst consisting of platinum alloyed with cobalt is thoroughly investigated via XANES and EXAFS, and it will be shown that XAS profits, among others, from the results of HRTEM and XRD to give substantial insight into some of the key characteristics – the strength of oxygen bondage – responsible for the catalysts' superior activity. After that, the main work of this thesis will be presented in chapter 6: The investigation of the practicability of site-selective XAS on a system of cobalt nanoparticles, which exhibit a variable coating resulting from a so-called "smooth oxidation" process [7]. It will be demonstrated that this promising technique of site-selective XAS can be further elaborated to establish a general strategy for the extraction of XAS spectra that describe the interior's and the coating's electronic, geometric and local atomic properties separately. For the Co nanoparticles in

1. Introduction

particular a core-shell–like structure is observed. The metallic core is determined to be crystalline in the hexagonal-close-packed phase with lattice constants only slightly smaller than for bulk Co. Further on, the shell (or coating) reveals itself to consist of only a few layers of mainly divalent cobalt-oxygen and cobalt-carbonate but without significant influence onto the interior. Finally, all results are summarized and conclusions, in view of implications for the current nanoparticle research as well as possible improvements, are discussed in chapter 7.

2. Theory of X-ray Absorption and X-ray Emission Spectroscopy

Nowadays a wide range of X-ray spectroscopic techniques such as X-ray photoelectron spectroscopy (XPS), energy-dispersive X-ray spectroscopy (EDX), X-ray diffraction (XRD), small angle X-ray scattering (SAXS), X-ray emission spectroscopy (XES), and X-ray absorption spectroscopy are available, each with different advantages and disadvantages. The focus of this thesis will be onto XAS and the related XES as well as onto combinations of those two, the resonant inelastic X-ray scattering (RIXS) and variations thereof. To perform valuable X-ray spectroscopic experiments an intense and tunable X-ray source is indispensable, which leads to the need for synchrotron radiation sources that are by now available all over the world. The electrons inside the synchrotron ring are forced to travel with constant velocity and on a circular trajectory for what reason they emit X-rays. These X-rays are - depending on the synchrotron ring - within the energy range from 0.1 to 100 keV, i.e. wavelength from 100 down to 0.1 Å, almost covering the hole range of atomic core level binding energies and interatomic distances, respectively.

2.1. Interactions of photons with matter

When a beam of X-rays, or more generally of photons, passes through matter it loses intensity due to its interaction with the matter. This loss is generally described by the Lambert-Beer law, which states that the attenuation A of light of energy E is linearly related to the properties of the material, i.e. to its density of absorbers N (per volume) and its thickness x:

$$A(E) = -\sigma(E) \, N \, x. \tag{2.1}$$

The proportionality constant $\sigma(E)$ is the energy dependent attenuation cross section (attenuation of a square unit), which in general consists of a coherent (Raleigh) and incoherent (Compton) scattering part as well as a photoelectric and pair-production part. This law comes along with a number of prerequisites (taken from [98]):

1. "The absorbers must act independently of each other";
2. "The absorbing medium must be homogeneously distributed in the interaction volume (realized by virtue of sample synthesis and/or preparation) and must not scatter the radiation" (automatically fulfilled for X-rays, *vide infra*);
3. "The incident radiation must consist of parallel rays, each traversing the same length in the absorbing medium" (ensured when using synchrotron rings as X-ray source);
4. "The incident radiation should preferably be monochromatic, or have at least a width that is more narrow than the absorbing transition" (a X-ray monochromator is always used in XAS and XES experiments); and

2. Theory of X-ray Absorption and X-ray Emission Spectroscopy

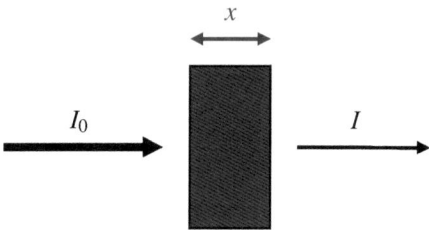

Figure 2.1.: Simple scheme of photon interaction with matter. Here x is the materials thickness and I_0 and I is the initial and transmitted intensity, respectively.

5. "The incident flux should only act as a non-invasive probe of the species under study. In particular, this implies that the light should not cause optical saturation or optical pumping, since such effects will deplete the lower level and possibly give rise to stimulated emission" (easily satisfied for transition metals under study).

If any of these conditions is not fulfilled, there will be deviations from Lambert-Beer's law.

Rewriting Eq. (2.1) in differential form, the attenuation dI of the total intensity I of photons travelling through matter is proportional to the thickness dx and to I:

$$dI = -\mu(E) \, I \, dx, \qquad (2.2)$$

where the linear energy dependent attenuation coefficient $\mu(E) = \sigma(E) \, N$ is introduced. Obviously, dI/I is just another (differential) description for the attenuation A. Integrating Eq. (2.2) yields

$$\ln(I/I_0) = -\mu x \qquad (2.3)$$

or

$$I = I_0 e^{-\mu x}, \qquad (2.4)$$

which is the standard form of the Lambert-Beer law and is visualized in Fig. 2.1. Here Eq. (2.3) is the one to be used in a XAS transmission experiment: The initial intensity I_0 and the transmitted intensity I are measured to obtain μx as a function of the X-ray energy E.

In Fig. 2.2 the mass attenuation coefficient (μ/ρ with density ρ) is plotted for cobalt (Co) over a wide energy range, so that all contributing parts are visible. At high energies (MeV) pair-production is relevant, which is the creation of an electron-positron pair from a photon in the electric field of the nucleus (or another electron). The latter is necessary in order to conserve the momentum of the initial photon. Consequently, the minimum energy for this process is two times the rest mass of the electron, i.e. 1.022 MeV, which makes it unimportant for X-ray spectroscopy. Another effect upon photon interaction with matter is scattering, i.e. the deflection of the photons from their initial trajectory due to collisions with electrons or the nucleus. This process can take place inelastically (incoherent- or Compton-scattering), where part of the photon energy is transferred to the scattered electron, which recoils and is ejected from its atom, or elastically (Coherent- or Rayleigh-scattering) where only the direction of the photon is changed. Coherent X-ray scattering is utilized in the course of a XRD experiment and yields a material specific diffraction pattern. As can be seen in Fig. 2.2 the coherent and incoherent scattering play a secondary role at energies below 100 keV.

2.1. Interactions of photons with matter

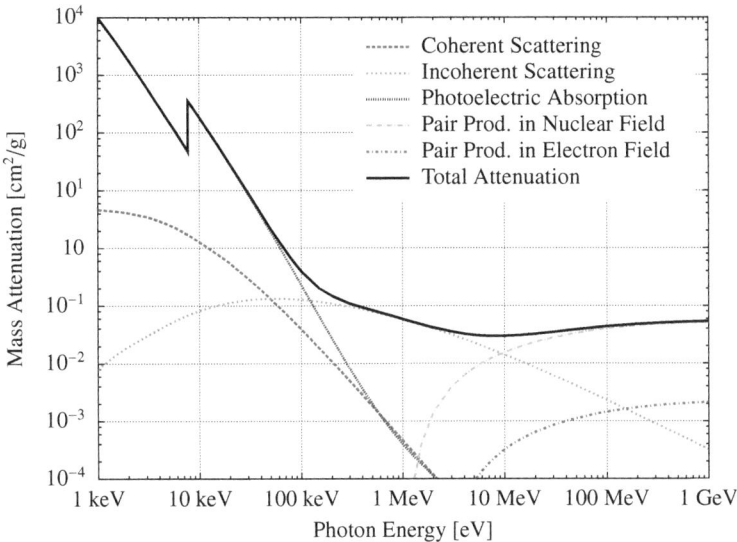

Figure 2.2.: Linear mass attenuation coefficient μ/ρ for cobalt on a double-logarithmic scale.

Finally, there remains one process that dominates the attenuation coefficient μ of X-rays, i.e. of photons in the energy range from 1 keV to about 100 keV, which is the photoelectric effect (see Fig. 2.2). Hereby the X-ray is "absorbed" by the atom or more precisely by a deep bound electron, which is ejected as a consequence, and therefore for XAS μ is called "absorption" coefficient. Each time the X-ray energy equals the binding energy of such a core electron, a sharp absorption edge appears that interrupts the otherwise continuous decrease of μ with increase of photon energy. In Fig. 2.2 such an edge appears at exactly 7.709 keV, where the deepest bound core-level electron of cobalt is excited, leaving behind a core hole. The detailed investigation of those edges to gain electronic and geometric information about a material is the focus of XAS. The process subsequently following such an X-ray absorption event is the relaxation of the excited atom by filling its core hole with a weaker bound electron, which is accompanied by the emission of a photon or Auger electron, carrying away the energy difference of these two atomic states. The detailed study of this fluorescent X-ray radiation (or alternatively of the Auger electrons) that reflects the electronic structure of an atom species is called XES (or Auger electron spectroscopy, AES). One can also make use of these emitted photons or electrons to gain a XAS spectrum, since the emission events are proportional to the absorption events. In this case the energy of the fluorescence radiation do not necessarily have to be highly resolved (however, it brings a lot of benefits as will be seen in section 2.3), one just has to count the emission events and the absorption coefficient is obtained as

$$\mu \propto I_f/I_0, \tag{2.5}$$

where I_f is the fluorescence intensity and I_0 the initial X-ray intensity.

2. Theory of X-ray Absorption and X-ray Emission Spectroscopy

Table 2.1.: Possible electron states for the case of the principal quantum number $n = 2$. l is the orbital angular momentum quantum number, j the total angular momentum, $2j+1$ the multiplicity and nl_j the term symbol. In the last column the completely filled states are shown with electrons symbolized by arrows (↑ = spin up, ↓ = spin down).

2	1	1/2, 3/2	2+4	$2p_{1/2}, 2p_{3/2}$	↑↓ ↑↓↑↓
	0	1/2	2	$2s_{1/2}$	↑↓
1	0	1/2	2	$1s_{1/2}$	↑↓
n	l	j	$2j+1$	nl_j	

2.2. X-ray absorption spectroscopy (XAS)

XAS is a widely-used technique for determining the local geometric and electronic structure of matter. It is applicable to any states of matter, i.e. solid, liquid or gaseous, as no particular long range order is necessary. It is element-specific, since the X-ray energy is tunable to an arbitrary edge of the element of interest. XAS is used in very different scientific fields including molecular and condensed matter physics, materials science and engineering, chemistry, earth science, and biology. The strong sensitivity to first neighbors makes XAS the tool of choice, in particular, for coordination chemistry and chemistry of catalysts and other nanostructures.

In a XAS experiment the X-ray energy is tuned by using a crystal monochromator [52] through an edge of the element of interest of the to be investigated material. These edges, arising from the ejection of deep bound core electrons, are labelled K, L, M, etc., which corresponds to the principal quantum number n = 1, 2, 3, ... of the main electron shell. The respective knocked out electrons are labelled 1s, 2(s,p), 3(s,p,d), etc., whereas the second (azimuthal) quantum number (s, p, d, ...) denotes the orbital angular momentum l = 1, 2, 3, ..., $n-1$. To characterize the orbitals and corresponding electrons completely, the magnetic quantum numbers $m_l = -l, ..., l$ (projection of the orbital angular momentum) and $m_s = \pm 1/2$ (projection of the intrinsic angular momentum, the spin s) are needed. From the Pauli exclusion principle that states that no two electrons within one atom can have the same set of quantum numbers n, l, m_l and m_s it follows furthermore: Each main shell n can contain

$$2 \sum_{l=1}^{n-1} \sum_{m_l=-l}^{l} (2l+1) = 2n^2$$

electrons.

However, in reality the angular momenta l and s are coupled due to the spin-orbit interaction, so that another quantum number j (the total angular momentum) is adequate to fully describe an electron state. It is the vectorial sum of l and s and can take the following range of values: $|l-s| \leq j \leq l+s$, which results in $2j+1$ electron states for each j. Since the electron spin s has the fixed value $1/2$, j can take the values, e.g., for n = 1, 2, as shown in Tab. 2.1.

The second last column of Table 2.1 shows the so-called term symbol nl_j for one electron systems, which has the general form for multi-electron systems $^{2S+1}L_J$, with S, L and J being the total spin, total orbital angular momentum and total angular momentum, respectively, and $2S+1$ the multiplicity (maximum number of electrons) of this state. In a relaxed atom the electron states are filled according to the Pauli principle up to the Fermi energy, which is the energy of the highest occupied electron state at absolute zero temperature. This is depictured in the last column of Tab. 2.1, where arrows symbolize the spin-up ($m_s = +1/2$) and -down ($m_s = -1/2$) electrons in each shell.

2.2. X-ray absorption spectroscopy (XAS)

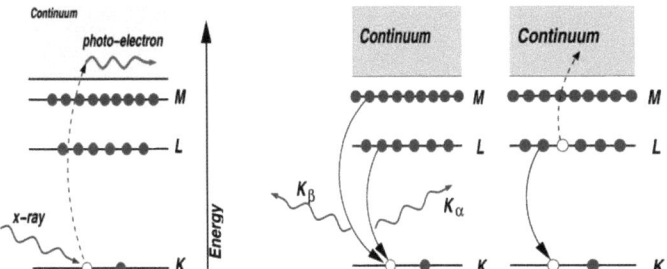

Figure 2.3.: The photoelectric effect where a X-ray photon is absorbed by a core-shell electron that is ejected, i.e. excited into the continuum (left). This is followed by the relaxation of the atom via X-ray fluorescence (K_α or K_β) (middle) or the Auger effect (right) (picture taken from [60].)

A typical example for a XAS event, i.e. the photoelectric effect, is shown schematically in Fig. 2.3. Here a X-ray photon hits a K-shell electron, which, as a result, is promoted to the continuum, that is to an unbound "state" outside the atoms binding range. This is followed by the atoms relaxation, where a 2p or 3p electron from the L- or M-shell re-fills the 1s core-hole. In this relaxation process the energy difference of the two electron levels is released in the form of radiation (i.e. the emission of a K_α or K_β photon, see Section 2.3 and Fig. 2.10 for more details regarding the nomenclature of the emitted photons), or via ejection of another (Auger) electron, whose kinetic energy is the difference of the released energy and the binding energy of the Auger electron. For X-ray absorption processes the Auger effect takes place mainly in the soft X-ray energy regime ($E_x < 2\,\text{keV}$) and diminishes for hard X-rays.

For a free atom this process would lead to a smooth absorption edge as was visible in Fig. 2.2 (and Fig. 2.6) and one could determine the type of atom from this (Moseley's law). However, since the focus is on solids, there is a huge number of atoms embedded in well ordered (or distorted) crystals or even amorphous structures. Due to this atomic neighborhood the region joining the absorption edge at the high energy side exhibits a lot of "fine structure". This X-ray absorption fine structure (XAFS) can be exploited to gain more sophisticated information about the material and will be explained in the next section.

2.2.1. X-ray absorption fine structure (XAFS)

In Fig. 2.4 a typical K-edge X-ray absorption spectrum for a transition metal is shown along with the standard division into three regions. As the energy of the X-ray photons is tuned through the edge (or binding) energy of the 1s electrons, the electrons are excited into the unoccupied orbitals. The most intense features arise, according to the quantum mechanical selection rules, due to electric-dipole transitions ($\Delta l = \pm 1$). Therefore the first features of the so-called "pre-edge", which are due to dipole-forbidden 1s→3d (or 4d, 5d) transitions, i.e. $\Delta l = 2$, are insignificant unless they are enhanced by overlapping p-orbitals of ligands or, when there is no inversion symmetry, of the metal itself, as is the case in Fig. 2.4. Such pre-edges only appear, at the K edge, for transition metals or compounds thereof, which have partially filled d orbitals. The pre-edge is followed by the strong rising edge, which originates in dipole allowed 1s→4p (or 5p, 6p) transitions. Both the pre-edge and edge region are summarized as near-edge X-ray fine structure (NEXAFS) or X-ray absorption near

2. Theory of X-ray Absorption and X-ray Emission Spectroscopy

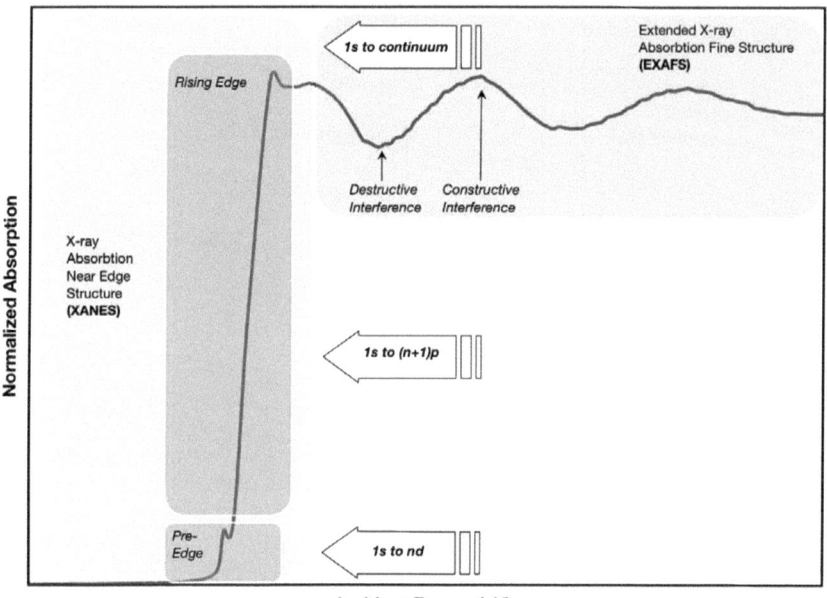

Figure 2.4.: Schematic K-edge X-ray absorption spectrum of a transition metal compound and its division into three regions. Dominant electron transitions (1s to $n = 3, 4, 5$) are assigned to each region (taken from [100]).

edge structure (XANES). The last and most far stretching region shows the extended X-ray absorption fine structure (EXAFS). It reaches up to several hundred eV above the edge. It arises mainly due to 1s→continuum transitions, i.e. due to electrons that are ejected from the atoms. Those photoelectrons are travelling through the material with a kinetic energy E_e equal to the excess energy $E_x - E_0$, where E_x is the X-ray energy and E_0 the (positive) binding or edge energy. It has to be pointed out here that the electron binding energies are precisely known, however, their position with respect to the XAS edge not. Nevertheless there are given standards how to define the position of the XAS edge. The most common choice, which is used in this work exclusively, is to take the maximum of the first derivative of an absorption spectrum (the first inflection point in the edge), and set it equal to the respective electron binding energy. Consequently, (electron) binding energy and edge energy is used synonymously. Here it is important to notice that the absolute energy scale is defined by this choice of the edge energy position. Thus, different XAS spectra are only comparable with respect to energies, if the edge energy is defined equally.

Actually the distinction between the XANES and EXAFS region is rather fluent and by no means strictly defined. Furthermore, it depends on the task and its aims how far, for example, the XANES region will be "extended". One borderline is given, as mentioned above, due to the final state of the electron transitions, whether it is a bound one or the continuum. Thereby, a borderline regarding the information that can be gained is given too: Electron structure due to probing of the unoccupied electron states (XANES) or geometric structure due to interaction with the atomic (and electronic)

2.2. X-ray absorption spectroscopy (XAS)

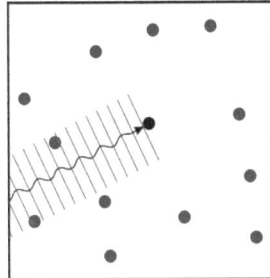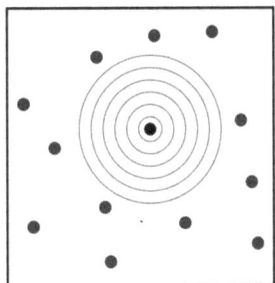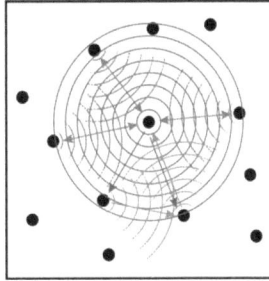

Figure 2.5.: The atom is hit by a X-ray (left). Hereby, a photoelectron is ejected in form of a spherical wave (middle) and interferes with its scattering part (right).

environment (EXAFS). Again, as was the case for the edge energy, this borderline can only be approximated with respect to a XAS spectrum, although it is theoretically well determinable for each element.

To understand why EXAFS, or the excitation of an electron to the continuum, provides geometrical information one switches from X-ray energy to the electrons (kinetic) energy ($E_e = \hbar^2 k_e^2$) or wave number

$$k_e = \sqrt{\frac{2m}{\hbar^2}(E_x - E_0)}, \tag{2.6}$$

which is a real number first for $E_x \geqslant E_0$, defining the minimum possible energy for EXAFS. According to de Broglie a wavelength $\lambda_e = 2\pi/k_e$ can be assigned to the photoelectron. Thus, as depicted in Fig. 2.5, the ejected photoelectron can be viewed as an outgoing spherical wave, which is scattered at the neigboring atoms, resulting in (partially) ingoing scattering electron waves. The in- and outgoing electron waves do interact, which results in an interference pattern that depends on the relation between λ_e and the distance to the neighboring atoms. The interference at the absorbing atoms position modulates the absorption probability μ, giving rise to an increase when the interference is constructive or a decrease when it is destructive (as will be explained later via Eq. 2.33). This oscillating fine structure is superposed onto the smooth energy dependence of a single atoms μ as is demonstrated in Fig. 2.6.

Here another possible distinction between XANES and EXAFS arises due to comparison of the photoelectrons' wavelength λ_e and the interatomic distance d. One can take the borderline between XANES and EXAFS as the point where $\lambda_e = a$. If one takes cobalt as an example again, which has in its hexagonal-close-packed structure (hcp) a Co–Co distance of $a = 2.51$ Å, the X-ray boundary energy would be $E_x = 7.733$ keV according to Eq. (2.6) with the Co K-edge energy $E_0 = 7709$ eV and $a = \lambda_e = 2\pi/k_e$. Here the XANES would extend to about 20 eV above the K-edge, however, the common choice is to set the borderline between XANES and EXAFS at 40 - 50 eV above the edge, as this region still shows sensitivity to the geometrical structure in E-space.

2.2.2. Multiple scattering formalism

The complete XAS regime can actually be uniformly described in terms of scattering waves, however, for this a more formal description of the process is necessary. X-ray absorption is the transition between two quantum states $|i\rangle$ and $|f\rangle$ with energies E_i and E_f, which can generally be described by

2. Theory of X-ray Absorption and X-ray Emission Spectroscopy

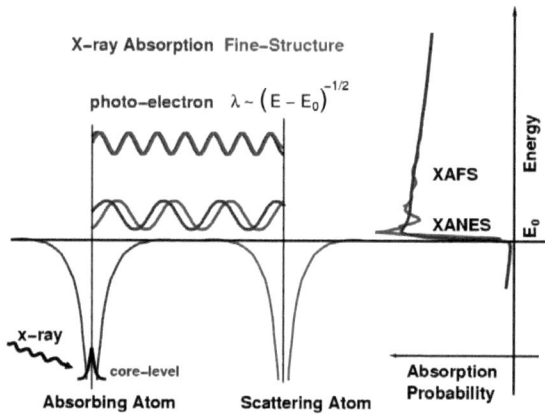

Figure 2.6.: XAS through the photoelectric effect for a single atom (blue lines only) and for an atom surrounded by other atoms (blue and red lines): The ejected photoelectron is scattered from a neighboring atom. The scattering electron wave, when returning to the absorbing atom, interferes with the initial photoelectron wave, which modulates the absorption probability (picture taken from [60]).

the use of time-dependent perturbation theory that leads to two alternative forms of Fermi's Golden Rule [32]:

$$\mu(E) \propto |\langle f|\mathcal{H}_{em}|i\rangle|^2 \, \rho(E_f) \tag{2.7}$$

$$\propto \sum_{f}^{E_f > E_F} |\langle f|\mathcal{H}_{em}|i\rangle|^2 \, \delta(E_x - E_i - E_f). \tag{2.8}$$

Here $\rho(E_f)$ is the density of unoccupied electron states at the final state energy E_f and $E = E_x - E_i$ is the photoelectron's energy. The initial state $|i\rangle$ consists of a X-ray photon, a core electron and no photoelectron, and the final state $|f\rangle$ has no X-ray, a core-hole and a photoelectron. The transition takes place by means of the electromagnetic interaction described by the Hamiltonian $\mathcal{H}_{em}$. In Eq. (2.8) one has to sum over all final states with energy E_f above the Fermi energy E_F (which is implied in $\rho(E_f)$), and the energy conservation is taking into account via the delta function. In order to solve Eq. (2.8) the one-electron approximation (for the initial and final state) is applied: No other electrons but the photoelectron are contributing, which is well fulfilled for deep 1s states, but gets worse for 2p etc., where multiplet effects [21] have to be taken into account.

The Hamiltonian $\mathcal{H}_{em}$ for the interaction of an electromagnetic field, mediated by the incident X-ray photon, with an electron of mass m has the general form

$$\mathcal{H}_{em} = \frac{1}{2m}(\vec{p} - e\vec{A})^2, \tag{2.9}$$

with the vector potential $\vec{A}(\vec{r}) = \vec{\epsilon} A_0 e^{i\vec{k}\cdot\vec{r}}$ that can be taken as a classical wave with polarization $\vec{\epsilon} \perp \vec{k}$. Since the electromagnetic field is rather weak, even for 3rd generation synchrotron sources, only the linear term in $\vec{A}$ (namely $\vec{p} \cdot e\vec{A}/2m$) of Eq. (2.9) has to be accounted for. Furthermore, one can in

2.2. X-ray absorption spectroscopy (XAS)

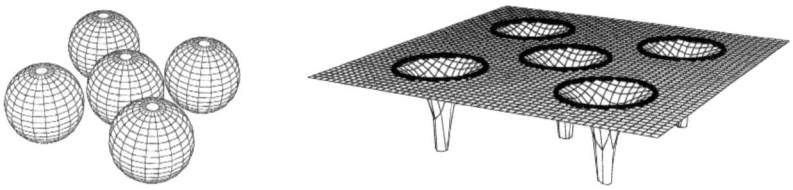

Figure 2.7.: Muffin-tin approximation: Non-overlapping spherical potentials separated by interstitial regions of constant potential (picture taken from [55]).

most cases only regard the dipole elements: $e^{i\vec{k}\cdot\vec{r}} = 1 + i\vec{k}\cdot\vec{r} - \frac{1}{2}(\vec{k}\cdot\vec{r})^2 + ... \simeq 1$. This is valid if $\vec{k}\cdot\vec{r} \ll 1$ or $|\vec{r}| \ll \lambda/2\pi$. For, e.g., the Co K-edge the X-ray wavelength would be: $\lambda/2\pi = \hbar c/E_{Co-K} \simeq 0.26$ Å and with the K-shell diameter estimated from the Bohr radius $a_0 = 0.53$ Å to $|\vec{r}| = 2 \cdot a_0/Z_{Co} \simeq 0.04$ Å. Hence, the dipole approximation is roughly fulfilled, nonetheless, the quadrupole etc. elements will be included in all calculations preventively. Lastly, one switches from momentum to position space by means of the commutator $[H, \vec{r}] = \vec{p}/m$ and gets

$$\mu(E) \propto \sum_{f}^{E_f > E_F} |\langle f|\mathcal{D}|i\rangle|^2 \, \delta(E_x - E_i - E_f), \tag{2.10}$$

with the dipole operator $\mathcal{D} = \vec{\epsilon} \cdot \vec{r}$.

In general there are two possibilities to solve Eq. (2.10). One could accurately describe both the initial $|i\rangle$ and the final state $|f\rangle$ in terms of molecular orbitals [4, 40] and solve the implied integral, or one could rewrite Eq. (2.10) to match the multiple scattering formalism, by using real space Green's functions as will be demonstrated in the following.

Assuming that all atom potentials v_i are (initially) non-overlapping spherical regions (which will be allowed to overlap in a self-consistent loop a posteriori) and that the interstitial space has a constant potential V_{int} (so-called Muffin-tin approximation, see Fig. 2.7), the propagation of the photoelectron is described by the Hamiltonian

$$H = H_0 + V, \tag{2.11}$$

with $V = V_{atom} + V_{int}$ and $V_{atom} = \sum_i v_i$ the total atomic potential. Following the well-known procedure now up to Eq. (eq:greens-fct-full) (see, e.g., chapter 7.1 of [79]) one is looking for the solution of $H|\psi\rangle = E|\psi\rangle$ or rearranging this:

$$(E - H_0)|\psi\rangle = V|\psi\rangle. \tag{2.12}$$

Since one is interested in elastic scattering of the photoelectron, $|\phi\rangle$ may be the solution for a freely propagating electron (i.e. $V = 0$) with the same energy E

$$(E - H_0)|\phi\rangle = 0. \tag{2.13}$$

Then Eq. (2.12) is obviously solved formally by the so-called Lippmann-Schwinger equation

$$|\psi\rangle = |\phi\rangle + \frac{1}{E - H_0 \pm i\eta} V|\psi\rangle, \tag{2.14}$$

with η infinitesimal small and $\eta \to 0^+$, to avoid a singularity. For $V \to 0$ Eq. (2.14) transforms to $|\psi\rangle =$

13

2. Theory of X-ray Absorption and X-ray Emission Spectroscopy

$|\phi\rangle$ as demanded. The Lippmann-Schwinger equation can also be solved formally by introducing the transition operator T, which will become important later

$$V|\psi\rangle = T|\phi\rangle. \qquad (2.15)$$

The kernel of Eq. (2.14) is identified as the Green's function describing the propagation of a free electron

$$G_0^{\pm} = \frac{1}{E - H_0 \pm i\eta}. \qquad (2.16)$$

The meaning of the superscript $\pm$ becomes clear upon determination of the precise expression for $G_0^{\pm}$ (see chapter 7.1 of [79]): $\pm$ stands for outgoing and incoming waves respectively and "$-$" will be omitted from no on, i.e. only the "normal" case that the photoelectron is escaping from the atom is regarded. The full electron propagator G is defined analogue by replacing H_0 with H from Eq. (2.11)

$$G = \frac{1}{E - H + i\eta}. \qquad (2.17)$$

Next a complete basis set $\sum_f |f\rangle\langle f| = 1$ is inserted into Eq. (2.17)

$$G = \sum_f \frac{|f\rangle\langle f|}{E - E_f + i\eta}, \qquad (2.18)$$

where $H|f\rangle = E_f|f\rangle$ is used. Now one needs to make use of the Sokhatsky-Weierstrass theorem

$$\lim_{\epsilon \to 0} \frac{1}{\omega \pm i\epsilon} = \mathcal{P}\frac{1}{\omega} \mp i\pi\delta(\omega)$$

($\mathcal{P}$ is Cauchy's principal value), whereof but the imaginary part is needed

$$\mathrm{Im}\left[\frac{1}{\omega \pm i\epsilon}\right]_{\epsilon \to 0} = \mp\pi\delta(\omega),$$

which applied to Eq. (2.18) leads to

$$\mathrm{Im}\,[G(E)] = \mathrm{Im}\left[\sum_f \frac{|f\rangle\langle f|}{E - E_f + i\eta}\right] = -\pi \sum_f |f\rangle\langle f|\,\delta(E - E_f). \qquad (2.19)$$

This equation will be plugged into the previous result for the golden rule Eq. (2.10) (remember that $E = E_x - E_i$):

$$\mu(E) \propto \sum_f^{E_f > E_F} \langle i|\mathcal{D}^\dagger|f\rangle\langle f|\mathcal{D}|i\rangle\,\delta(E - E_f) \qquad (2.20)$$

$$\propto -\frac{1}{\pi}\langle i|\mathcal{D}^\dagger\,\mathrm{Im}\,[G(E)]\,\mathcal{D}|i\rangle\,\Theta(E - E_F), \qquad (2.21)$$

where in the last step the Heaviside step function Θ appears, which assures that $\mu(E) \neq 0$ only for $E > E_F$, which was guaranteed before in Eq. (2.20) via the sum and the δ-function.

Finally, in Eq. (2.21), the sum of final states is rewritten in terms of the full Green's function G. It remains to be shown, how the absorption coefficient μ in Eq. (2.21) is calculated and how one can

2.2. X-ray absorption spectroscopy (XAS)

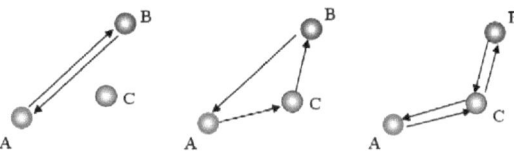

Figure 2.8.: Visualization of single-, double-, and triple-scattering paths for three fixed atoms, as described by the full Green's function Eq. (2.25).

visualize this Green's function formalism. For this purpose Eq. (2.17) will be rearranged into

$$G = (G_0^{-1} - V)^{-1} = (1 - G_0 V)^{-1} G_0 \qquad (2.22)$$
$$\Longleftrightarrow \quad G = G_0 + G_0 V G \qquad (2.23)$$
$$\Longleftrightarrow \quad G = G_0 + G_0 T G_0 \qquad (2.24)$$

by performing some operator algebra and using Eq. (2.15) in the last step. If one iterates the first of the two Dyson equations (Eq. (2.23))

$$G = G_0 + G_0 V G_0 + G_0 V G_0 V G_0 + \ldots \qquad (2.25)$$

one eventually can understand the meaning of G: The first term G_0 describes the propagation of a free electron between two points in an arbitrary space. The second term describes all single scattering events at one atom with potential v_i (recall that $V = \sum_i v_i$ for standard case that $V_{int} = 0$), the third term describes all double scattering events and so forth (compare Fig. 2.8). G therefore describes all possible paths that the photoelectron can scatter from an arbitrary number of the surrounding atoms before the core-hole is refilled. Here it is important that only closed paths are contributing to the XAS absorption coefficient, i.e. only those that begin and end at the absorbing atom.

The explicit calculation of the full Green's function depends on the energy range, whether one is dealing with XANES (small energies) or EXAFS (high energies). In the latter case one can just use Eq. (2.25) and sum up the multiple scattering paths until they converge, which occurs quite fast, even double-scattering is in most cases an order of magnitude smaller than the single scattering events. For small energies, however, no convergence can be achieved due to the large photoelectron wavelength λ_e, relatively to the nearest neighbour distance d, which amplifies the importance of the multiple scattering events. Thus, for XANES Eq. (2.22) has to be computed explicitly or, instead of the full potential V, the single site scattering matrices t, defined via the transition operator (recall Eq. (2.15) $T = t + tG_0 t + tG_0 tG_0 t + \ldots$ are calculated (for explicit derivation see [5]) and references therein). This leads to an expression for G similar to Eq. (2.25)

$$G = G_0 + G_0 t G_0 + G_0 t G_0 t G_0 + \ldots \qquad (2.26)$$

and hence one similar to Eq. (2.22):

$$G = (1 - G_0 t)^{-1} G_0. \qquad (2.27)$$

G_0 as well as t can be cast into matrix form by using an appropriate basis set (angular-momentum representation), which allows for explicit calculations, limited by computer power only [3, 71].

One now may ask why, given such a strong tool for accurate calculations of a XAS measurement, not even the EXAFS is calculated exactly via Eq. (2.27), but with a conveniently truncated path expansion Eq. (2.26). The reasons are that (compare [74]): (1) it is not possible to include easily physics that grow important first at energies above the XANES region (e.g. thermal and static disorder, core-hole potential); (2) accurate calculations would provide an unnecessarily high degree of detail about the electronic structure that will be obscured by the various damping factors and lifetime broadenings; (3) the number of basis set functions is a rapidly increasing function of energy above the edge, which becomes incomputable. Furthermore, it would be impossible to disentangle the different paths and extract the specific information they give, like bond-length and -angle for each shell. Thus, for EXAFS it is reasonable to stick to the path expansion Eq. (2.26), which has to be elaborated further, however, so that it is capable of describing a real measurement exhaustively.

2.2.3. The EXAFS equation

For the investigation of the EXAFS it is convenient to separate the Green's function G and the absorption coefficient μ, respectively, into a part describing the central absorbing atom G^c and one describing the scattering at the neighborhood G^{sc}:

$$G = G^c + G^{sc}, \quad (2.28)$$

where $G^c = G_0$ and $G^{sc} = G_0 t G_0 + G_0 t G_0 t G_0 + ...$ (compare Eq. (2.26)). If one now assumes the general structure for the absorption coefficient

$$\mu(E) = \mu_0(E)[1 + \chi(E)], \quad (2.29)$$

where μ_0 is the absorption of an atomic like state and χ incorporates the XAFS, and compares this with Eq. (2.21), keeping Eq. (2.28) in mind, it follows

$$\chi \propto Im(G^{sc}) \simeq Im(G_0 t G_0). \quad (2.30)$$

Here, furthermore, Born's approximation is applied, which is valid when the effect of the scatterer is weak and thus $|\psi\rangle \to |\phi\rangle$ on the right-hand side of Eq. (2.14). This leads to the replacement $G \to G_0$ on the right-hand side of Eq. (2.23), which is equal to truncating Eq. (2.25) after the single-scattering term, and thus, only $G_0 t G_0$ appears in Eq. (2.30). Actually, Born's approximation is not valid anymore the closer one gets to the XANES region, i.e. the smaller the energy becomes. However, in Ref. [72] it is demonstrated that Born's approximation can be undone while preserving the (form of) the final result (*vide infra*).

In the following the EXAFS equation will be qualitatively derived. First, one switches to k-space according to Eq. (2.6), with k being the photoelectrons wave vector. Further on, spherical waves are assumed for the outgoing electron and its backscattered part, described by G_0, and an explicit expression for the scattering operator t is used:

$$\chi(k) \propto Im\left[\frac{e^{ikR}}{kR} k f(k) e^{i\phi(k)} \frac{e^{ikR}}{kR}\right] \quad (2.31)$$

$$\chi(k) \propto \frac{f(k)}{kR^2} sin[2kR + \phi(k)]. \quad (2.32)$$

R is the distance to the first neighbor and $f(k)$ (scattering amplitude) as well as $\phi(k)$ (scattering phase shift) are the scattering properties of the neighboring atom. Both scattering quantities depend on Z

which makes EXAFS sensitive to the species of the atomic neighborhood.

Now corrections terms will be added step-by-step to Eq. (2.32) which is still too oversimplified, so that real EXAFS measurement can be described:

- There is rarely only one pair of atoms (absorber and scatterer), but a number N of identical atoms with in the same distance R. These atoms within the same distance are called coordination shell and consequently N is the coordination number.

- Furthermore, even these identical atoms are in general somewhat displaced in the crystal structure by static and/or thermal disorder, which is accounted for by a Debye-Waller factor $e^{-2k^2\sigma^2}$ with $\sigma^2 = \sigma_{sta}^2 + \sigma_{the}^2$.

- Until now the fact was neglected that the photoelectron may also scatter inelastically. Additionally it has to return to the absorbing atom before the core-hole is refilled by another electron, i.e. the core-hole lifetime has to be regarded. Both effects will be included in a damping factor $e^{-2R/\lambda(k)}$. Here, λ is the mean-free-path of the photoelectron, that is the distance before one of the two effects occur.

- The whole process has been treated as a one-electron event, which was a crucial assumption, but of course other electrons could be excited during a excitation event too and lead to the well-known shakeup and shakeoff processes. These can be accounted for by a constant amplitude-reduction factor S_0^2 [74].

- Lastly, real systems do not inhere atoms in only one distance R and not only one type of atom. This is easily accommodated for, by just summing up the contributions from each atom type j with coordination number N_j and distance R_j.

Adding all these corrections to the initial Eq. (2.32), one ends up with the well-known EXAFS equation [80]:

$$\chi(k) = S_0^2 \sum_j \frac{N_j f_j(k)}{kR_j^2} e^{-2k^2\sigma^2} e^{-2R_j/\lambda(k)} \sin[2kR_j + \phi_j(k)]. \tag{2.33}$$

It describes the EXAFS by a summation of all single-scattering paths that belong to the various coordination shells j of the (different) atom species. It has been shown in Ref. [72] that this equation is also valid upon inclusion of all possible multiple scattering paths (as aforementioned), by "just" replacing the scattering amplitude f by an effective one f_{eff}, wherein all modifications are included. Further information regarding Eq. (2.33) are gathered in appendix D.

2.3. X-ray emission spectroscopy (XES)

Unlike XAS, X-ray emission spectroscopy (XES), or X-ray Raman scattering as it is often called, is a second order optical process, since prior to the emission event the atom of interest has to be excited. The term "scattering" stems from the fact that the whole process - an incoming X-ray is absorbed and subsequently a X-ray with distinct energy is emitted - can also be viewed as an inelastic scattering event of one X-ray. If the excitation energy is tuned through a resonance, e.g. 1s or 2p core states, the process is called resonant inelastic X-ray scattering (RIXS). The excitation takes place due to absorption of an X-ray, as was explained in detail in the last section. However, if the excitation energy exceeds the binding energy of the deepest bound states (K-shell), one is dealing with non-resonant inelastic X-ray scattering (NRIXS), which is independent of the absorption process, since

2. Theory of X-ray Absorption and X-ray Emission Spectroscopy

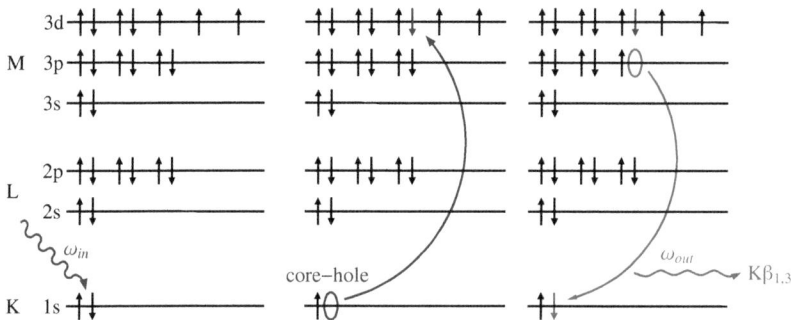

Figure 2.9.: 2nd order optical process as described by the Kramers-Heisenberg equation Eq. (2.34) for the case of cobalt. Co has the ground state configuration [Ar]$3d^7 4s^2$ ($4s^2$ electrons not drawn) or short: $|g\rangle = 3d^7$ + X-ray. The intermediate state is $|n\rangle = 1s3d^8$ (1s stands for the core-hole) and the final state is $|f\rangle = 3p^5 3d^8$ + X-ray.

no new intermediate electron state is created that could interact with the final state (see Eq. 2.34). Nonetheless, it should be kept in mind that the emission process occurs in the presence of the core-hole left over by the absorption. The case of RIXS and NRIXS, where the energy of the incident X-rays is tuned through an absorption edge and above, is theoretically described by the the following part of the Kramers-Heisenberg formula [46, 47]

$$I(\omega_{in}, \omega_{out}) = \sum_f \left| \sum_n \frac{\langle f|T_2|n\rangle\langle n|T_1|g\rangle}{E_g + \omega_{in} - E_n - i\Gamma_n/2} \right|^2 \times \delta(E_g + \omega_{in} - E_f - \omega_{out}), \quad (2.34)$$

Here ω_{in} is the energy of the incoming X-ray that excites the atom, by means of the transition operator T_1, from the ground state g with energy E_g to the intermediate state n with energy E_n. Γ_n is the lifetime broadening due to the core-hole in the intermediate state. The subsequent radiative decay of the atom, where a X-ray of energy ω_{out} is emitted, is described by T_2 and leads to the final atomic state f with energy E_f. Both T_1 and T_2 are equivalent to $\mathcal{H}_{em}$ (introduced in Eq. 2.8), i.e. they describe the electromagnetic interaction in the form of dipole, quadrupole, etc. elements. The summation is with respect to all intermediate and final states n and f, respectively, that are accessible via the specific in- and outgoing X-ray.

In Fig. 2.9 one special case for Co is drawn: 1s→3d followed by 3p→1s, whereas the latter transition is accompanied by the $K\beta_{1,3}$ X-ray emission. Of course, the intermediate state could also be the 4p levels or the continuum and the decay could alternatively proceed via the more probable 2p→1s transition, i.e. $K\alpha_{1,2}$ emission (see Fig. 2.10 for a scheme of XES transitions). Obviously, X-ray absorption probes the unoccupied electron states while X-ray emission is complementary as it probes the occupied ones, and in the case of resonant excitation both processes are strongly coupled. This coupling is visible in Eq. (2.34), due to the interference terms that appear upon squaring the sum of matrix elements with identical final but different intermediate states.

A general measurement (detailed description in section 3.3 and 3.4) of such a resonant second-order optical process yields a 3 dimensional RIXS plane, which is shown for cobalt(II)-Oxide (CoO) in Fig. 2.11 (top-right) as a contour plot. Each intensity point there depends on excitation (ω_{in})

2.3. X-ray emission spectroscopy (XES)

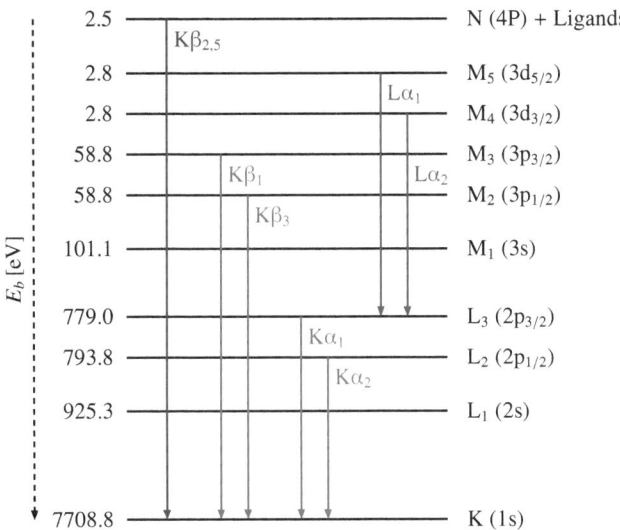

Figure 2.10.: Schematized XES transitions for Co that are appearing in this work. On the left is given the electron binding energy E_b (taken from the NIST database [61]) and on the right the edge label and in brackets the corresponding electron orbital.

and emission energy (ω_{out}). This allows for the extraction of emission spectra in dependence of fixed excitation energy, leading to highly resolved (depending on the spectrometer) resonant X-ray emission (RXES) (Fig. 2.11, left), or of excitation spectra (XAS) in dependence of appropriately fixed emission energy, which leads to selective high-resolution fluorescence-detected XAS (HRFD-XAS) (Fig. 2.11, bottom). Note that sometimes in the literature (N)RXES and (N)RIXS are used synonymously in contrast to the usage in this work. (N)RIXS will be strictly used when dealing with a three dimensional plane and (N)RXES for the case of a two dimensional emission spectrum. It should be mentioned that both HRFD-XAS and (N)RXES spectra can also be obtained without measuring a whole RIXS plane. The former, upon recording a certain fluorescence energy only while tuning the incident energy and the latter, upon fixing the incident energy and scanning the fluorescence (see also section 3.4).

2.3.1. High-resolution fluorescence detected XAS (HRFD-XAS)

The reason for the high-resolution of XAS spectra, that are extractable from RIXS planes, can be understood upon visualization of the core-hole lifetime broadenings as is done in Fig. 2.12. The top picture shows a section of a CoO RIXS plane with axes ω_{in} and ω_{out}. The edge region is shown and the pre-edge features of CoO are visible as nearly isolated resonances. At the bottom picture in contrast ω_{out} is replaced by the energy transfer, which is just the difference $\omega_{in} - \omega_{out}$, i.e. the final-state energy (as ω_{in} is the intermediate state energy), and which resulted in, roughly speaking, a 45° rotation of the RIXS plane. The latter representation of RIXS data is the more common and more convenient one, since here the core-hole lifetime broadenings of the intermediate state (Γ_{1s}) and the final state (Γ_{3p}) are parallel to the axes (compare Ref. [27]). In other words, when going along the excitation energy

2. Theory of X-ray Absorption and X-ray Emission Spectroscopy

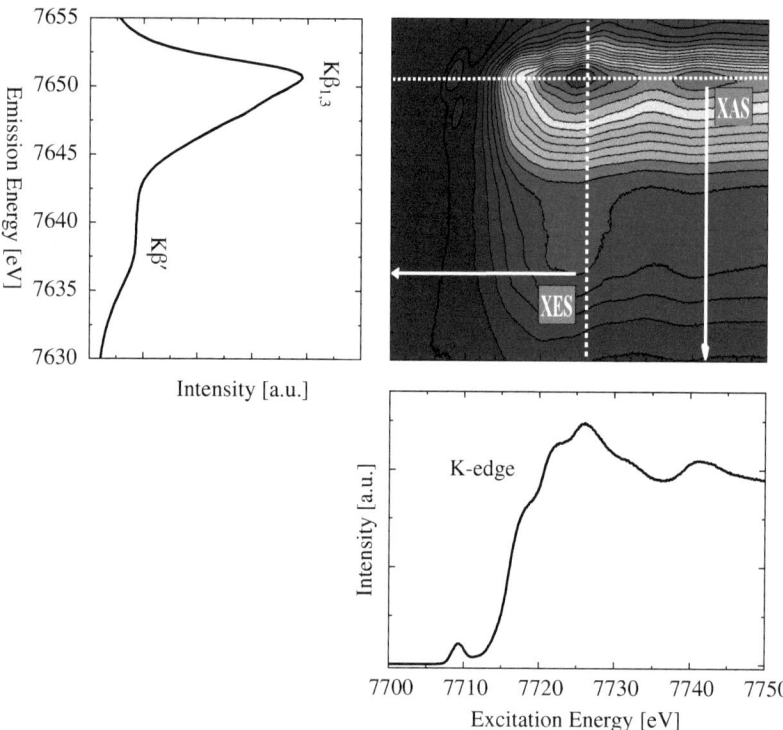

Figure 2.11.: 1s3p RIXS plane of CoO (measured at ID26, ESRF, see section 6.3) (top-right). High-resolution Co K$\beta_{1,3}$ RXES spectrum extracted at excitation energy of 7726.0 eV on the left (vertical white dashed line in RIXS plane) and Co K-edge HRFD-XANES spectrum extracted at emission energy of 7650.8 eV at the bottom (horizontal white dashed line in RIXS plane).

axis in Fig. 2.12 (bottom), each point is broadened by the 1s core-hole, but not by the 3p core-hole and vice-versa when going along the emission energy axis. However, a HRFD-XANES spectrum is extracted at fixed emission energy and, in the respective representation (Fig. 2.12, top), Γ_{1s} stretches diagonally through the plane, since this (original) RIXS plane is rotated about, roughly speaking, −45° relative to Fig. 2.12 (top). Consequently, a XAS spectrum at fixed emission energy of 7651 eV, visible as a horizontal line with label $I(\omega_{in}, \omega_{out}=7651\,\text{eV})$ in Fig. 2.12 (top), is not influenced by the full Γ_{1s} lifetime broadening as it would be in a classical XAS measurement. To be precise, it is influenced partly by Γ_{1s} and by Γ_{3p}. It can be shown [20] that the effective lifetime broadening in HRFD-XAS is $1/\sqrt{(1/\Gamma_{1s}^2 + 1/\Gamma_{3p}^2)}$, which in the current case, with $\Gamma_{1s} = 1.33$ eV (taken from [48]) and $\Gamma_{3p} = 1.189$ eV (calculated by FEFF [75]) yields $\Gamma_{\text{effective}} \simeq 0.8$ eV. One has to keep in mind though, the spectral broadenings due to the monochromator and the spectrometer too. The former affects the absorption features, i.e. broadens along the direction of the HRFD-XAS extraction, and the latter affects the emission features, i.e. intensifies Γ_{3p}. The experimental energy resolutions of the

2.3. X-ray emission spectroscopy (XES)

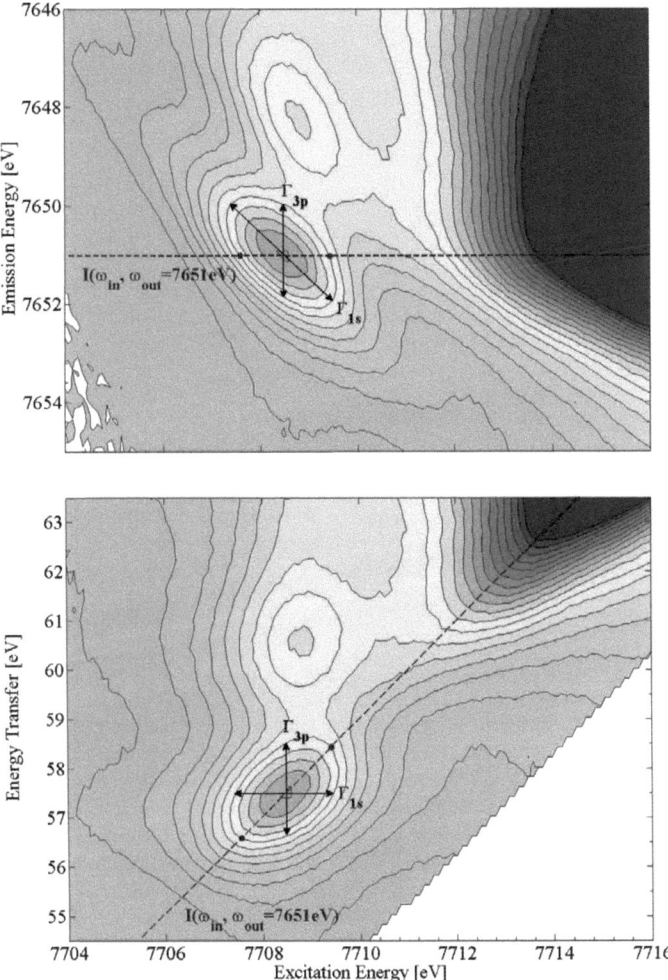

Figure 2.12.: 1s-3p RIXS plane of CoO. Also shown are the orientations of the core-hole lifetime broadenings Γ_{1s} and Γ_{3p} as well as the direction for extraction of a HRFD-XAS spectrum, labelled $I(\omega_{in}, \omega_{out}=7651\,\text{eV})$. Top: Emission against excitation energy is shown. Bottom: Energy transfer against excitation energy is shown.

2. Theory of X-ray Absorption and X-ray Emission Spectroscopy

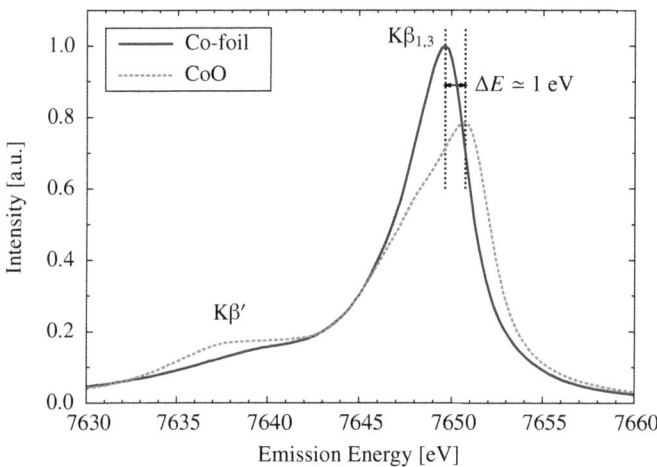

Figure 2.13.: Normalized $K\beta_{1,3}$ emission line of Co-foil and CoO (as powder). ΔE denotes the chemical shift due to different valencies of the two shown samples.

devices utilized for the spectra in Fig. 2.12 have been $\simeq 1\,\text{eV}$ for the monochromator and spectrometer likewise, i.e. similar to the lifetime broadenings which is a crucial point. For an example with precise calculations of effective spectral broadening with respect to core-hole lifetimes and instruments see [89].

2.3.2. Selectivity of HRFD-XAS

The origin of the "selectivity" of HRFD-XAS spectra is found in the chemical sensitivity of emission lines, which is connected, in the case of transition metals like Co, to the 3d valence band. This valence band overlaps with the initial state of the emission process, i.e. 3p in the case of $K\beta_{1,3}$ emission, resulting in a 3p-3d exchange interaction. The $K\alpha_{1,2}$ emission, which has a 2p initial state that sparsely overlaps with 3d, is less sensitive hence. Emission starting from the 3d level itself, or even from higher valence states (4p or ligand orbitals), are most sensitive albeit considerably less intense.

In Fig. 2.13 the $K\beta$ spectra (normalized to the spectral area) of a 6 μm thick metallic cobalt foil (Co-foil) and a cobalt(II)-Oxide powder (CoO) are shown. The peak at 7649.4 eV is the $K\beta_{1,3}$ line, arising from 3p→1s electron transitions - or more precisely: $3p_{1/2}$ and $3p_{3/2}$ to 1s, separated by about 1 eV see Fig. 2.10 by spin-orbit interaction. Another broad less intense peak appears at lower energies (about 7637 eV), labelled $K\beta'$. The splitting into $K\beta_{1,3}$ and $K\beta'$ originates in the strong 3p-3d exchange interaction, i.e. the orientation of the 3p hole spin relative to the unpaired electron spins in the 3d shell [25]. Since those unpaired electron spins preferably align themselves upwards according to Hund's rules, the promotion of a 1s spin-down electron to the 3d-level is preferable (compare Fig. 2.9). Consequently, the 3p→1s transition of spin-down electrons is reflected by the main peak $K\beta_{1,3}$, and the spin-up transitions mainly give rise to $K\beta'$. Since the number of unpaired 3d electrons is increasing from metallic to oxidized Co, this effect is stronger in CoO compared to Co. Obviously,

2.3. X-ray emission spectroscopy (XES)

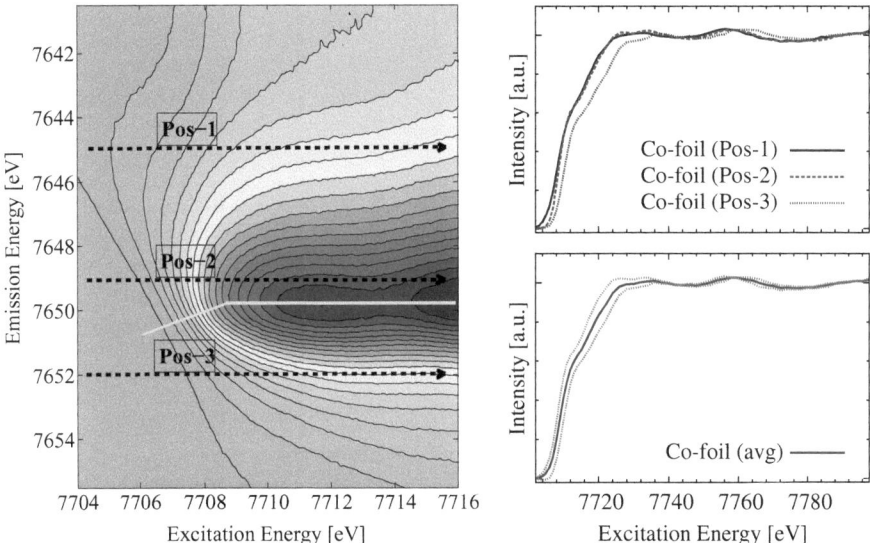

Figure 2.14.: Left: Section of the 1s3p RIXS plane of Co-foil. Emission against excitation energy is shown. The arrows indicate the direction where the HRFD-XANES spectra are extracted from. Right-top: The formerly described (normalized) HRFD-XANES spectra. Right-bottom: Co-foil HRFD-XANES spectrum with standard deviation, averaged from the three Co-foil spectra shown in the upper figure.

with decreasing net valence spin the two peaks are moving towards each other, and eventually $K\beta'$ merges into $K\beta_{1,3}$, as is almost the case for Co-foil in Fig. 2.13.

By virtue of the 3p-3d exchange interaction two selective modes arise for HRFD-XAS:

(1) By extracting HRFD-XAS spectra at the peaks, $K\beta_{1,3}$ and $K\beta'$, spin-selectivity is possible. However, the spectra are only partially selective since the pure spin-up and spin-down spectra are overlapping, at least in the $K\beta_{1,3}$ peak region (compare work on Fe^{2+} in Ref. [20], chapter IV.C]).

(2) The $K\beta_{1,3}$ peak of CoO is shifted about 1 eV to higher energies with respect to Co as denoted in Fig. 2.13. For mixed valence compounds hence, one can extract HRFD-XAS spectra, e.g., at the peak positions of the different valence compounds that are present, to gain valency-selective spectra, however, only "partially" selective since the peaks are overlapping again.

The latter mode - partial valency-selectivity - will be further elaborated in chapter 6 to achieve pure valency-selectivity.

2.3.3. Lifetime influence onto HRFD-XAS

In section 2.3.1 it was demonstrated how HRFD-XANES benefits from the diagonally stretching Γ_{1s} lifetime broadening that results in a reduced effective broadening. However, this benefit is likewise a

2. Theory of X-ray Absorption and X-ray Emission Spectroscopy

drawback, when several HRFD-XANES spectra are to be investigated that are extracted from distinct emission energies. To make this clear, a section of the RIXS plane of cobalt metal (Co-foil) is shown on the left of Fig. 2.14. The excitation energy is plotted against the emission energy and the development of the $K\beta_{1,3}$ peak from small to large excitation energies is visible. Obviously the contour lines are running obliquely, with respect to both axes in the edge region, until about 7709 eV denoted by the grey diagonal line and become approximately symmetric with respect to the line of maximum intensity (grey horizontal line), i.e. the Raman shift. This behavior leads to differences when measuring $K\beta_{1,3}$ detected spectra at distinct constant emission energies, even for a mono-metallic system as is shown in the right panel of Fig. 2.14. These three normalized (according to appendix C) HRFD-XANES spectra of a Co foil are extracted along the directions marked by horizontal arrows in the left-panel of Fig. 2.14. In these HRFD-XANES spectra one can observe significant variations with respect to the intensity of features within the edge (at 7709 eV) and the white line (around 7730 eV). Furthermore, the edge position is energetically shifted for Co-foil at Pos-1 about -1 eV and for Co-foil at Pos-3 about +1 eV, relative to Co-foil at Pos-2 which exhibits the correct edge position of 7709 eV.

The origin for this appearance of the RIXS plane, and the resulting differences in the HRFD-XANES spectra, is the orientation of the different core-hole lifetime broadenings, as was explained in section 2.3.1 (see Fig. 2.12). The HRFD-XANES spectra are extracted at fixed emission energy - the straight horizontal arrows in Fig. 2.14 - but the lifetime broadening of the 1s core-hole is stretching diagonally through this plane, i.e. each resonance, whether an isolated one or the continuum states (which can be understood as an infinite number of resonances), influences other states at completely different diagonally shifted positions. This has to be considered when interpreting resonances in two-dimensional spectra extracted from a RIXS plane and is the reason for the slanting contour lines in the edge region of a RIXS plane (when plotting emission vs. excitation energy) and leads to the differences that are visible in the HRFD-XANES spectra of Fig. 2.14.

Obviously, the most reliable HRFD-XANES spectrum is the one recorded at the fluorescence peak position (a horizontal line at 7649 eV in Fig. 2.14), as it exhibits the strongest signal and therefore is lest disturbed by lifetime broadenings of other "resonances". In order to get a pure valency-selective XANES spectrum, however (which is a main task of this work), several HRFD-XANES spectra are necessary (see chapter 6), and thus the differences, visible in the left panel of Fig. 2.14, have to be taken into account. Here it is advantageous when one of the positions for HRFD is (about) the fluorescence peak and the others are located on both lower and higher energies relative to it, because then the average spectrum is most close to the one recorded at the peak itself, as the differences are almost cancelled out. As a consequence, a pure valency-selective XANES spectrum will strongly depend on the chosen positions. To have a basis for estimating the validity of a determined pure valency-selective spectrum, the averaged HRFD-XANES spectra of its homovalent compounds (e.g. zerovalent Co-metal and divalent Co-oxide) can be compared to the respective HRFD-XANES spectrum recorded at the peak.

It has to be noted that the exact solution would be to deconvolute each data point in the RIXS plane with respect to all broadenings (instrumental and lifetime), but is beyond the scope of this work. Of further importance is that, when restricting the analysis to the EXAFS region, these differences are negligible since they are vanishingly small compared to the strong oscillations of the fine structure, and since the energy shifts are removed already by the transformation to k-space (remember section 2.2.1 and Eq. 2.6).

3. Experiment

In this chapter the principles of a synchrotron radiation facilities, as well as the respective beamlines where the X-ray spectroscopic measurements of this work have been performed, will be presented. These are, in Germany, the "Angströmquelle-Karlsruhe" (ANKA) at the "Karlsruhe Institute of Technology" (KIT) in Karlsruhe, as well as the "Double ring storage" (DORIS III) at the "Hamburger-Synchrotronstrahlungs-Labor" (HASYLAB) which is part of the "Deutsches-Elektronen-Synchrotron" (DESY) in Hamburg and lastly, in Grenoble, France, the "European Synchrotron Radiation Facility" (ESRF). Their main characteristics are listed in Table 3.1.

3.1. The synchrotron radiation facility

A synchrotron radiation facility, or synchrotron light source, is a specific type of particle accelerator whose aim is to accumulate a high electron current and maintain it for a long lifetime. Thus, in contrast to a standard particle accelerator, that, when the maximum energy is reached, lets its particles collide so that all produced energy is released in one blink of an eye, the synchrotron light source is specialized to conserve this (current of) electrons with a certain (maximum) energy as long and stable as possible. This is realized upon carefully synchronizing the magnetic fields, that force the particles onto the circular trajectory, and the electric field, that accelerates the particles, with the travelling particle beam (compare [107]).

In Fig. 3.1 a scheme for a typical synchrotron radiation facility is shown. In the center (1) the electrons are produced by an electron gun via thermoionic emission and are immediately attracted by a positive potential applied behind the gun. Here, they already gain an energy of about 100 keV and are put into bunches. Subsequently these bunches enter the linear accelerator (linac) (2), where they travel through several radio-frequency (RF) cavities (working at some GHz). Here the electron bunches successively gain an energy of some 100 MeV, since the length of the cavities is customized to the electrons movement so that they are always attracted. The pre-accelerated electron bunches are then fed into the booster ring (3) – a small synchrotron ring – where they reach their final energy of some GeV, by means of one RF cavity that provides an alternating electrical field with frequency of several hundred MHz, and that is synchronised with magnets for focusing and bending. The strength of both the electric and magnetic field is increased as the electrons gain kinetic energy, so that their path can be kept constant while they are increasingly accelerated. As the electron beam reaches the end-energy, it is injected into the storage ring (4), where its energy loss due to emission of synchrotron radiation is compensated by RF cavities. Inside the storage ring the electron beam is focused by quadrupole magnets and "bended" by dipole magnets (green and red devices in Fig. 3.1) in the arcs, to ensure the circular trajectory. In between the arcs are straight sections that can host insertion devices like "wigglers" or "undulators", and which are besides the bending magnets the sources of synchrotron radiation, which is highly focused along the moving direction of the electrons into a forward cone, due to the electrons' relativistic speed. The emitted synchrotron radiation eventually is utilized in so called "beamlines" (5), where the diverse measurements are performed.

Inside the linac, booster ring and storage ring, a ultra-high vacuum of typically about 10^{-13} bar is

3. Experiment

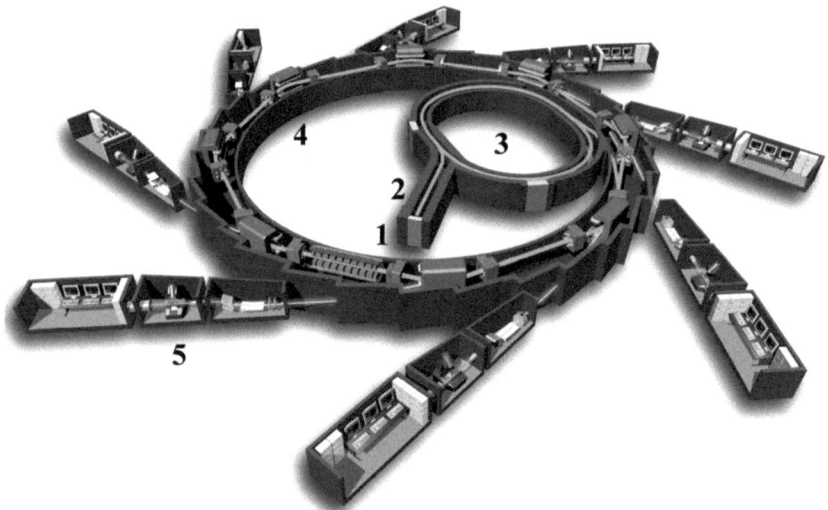

Figure 3.1.: Typical scheme of a synchrotron radiation facility (taken from [107]), with its different stations labelled and explained in the text.

required, since all collisions with gas particles will decrease the electron current and its energy. This is even more important for the storage ring, as the lifetime of the electron beam is dependent on there being almost no collisions. Consequently, owing to the strong focusing, a relatively thin (in diameter) stainless steel tube is used for the electrons to travel, which can be evacuated easily as is demanded.

Wigglers and undulators that are build into the straight sections of the storage ring, are devices that contain a series of bending (dipole) magnets of alternating polarity, through which the electrons have to travel as shown in Fig 3.2 (a) and (b). Due to the alternating attraction of the magnets, the electrons are "wiggling" (or "undulating") along a straight path, thereby emitting synchrotron radiation like in the bending magnets of the storage ring, though more intense. For a number N of dipole magnets inside the wiggler or undulator, the synchrotron intensity is enhanced by N for the former and by N^2 for the latter. The main difference now between a wiggler and an undulator is the strength of the dipole magnets. It is higher for a wiggler, which leads to larger deflections of the electrons (and higher X-ray energy), and thus the emitted cones have a large angle towards the central axis and do not overlap, so that the resulting radiation covers a wide spectral range. The small deflections inside the undulator in contrast, lead to the overlap and interference of all emission cones and consequently to most intense synchrotron radiation with a narrow bandwidth, i.e. high brilliance, as can be seen in Fig. 3.2 (c) (see [108] for more details about synchrotron radiation facilities and its devices).

At each synchrotron facility one beamline was visited for the X-ray measurements, each dedicated to a specific purpose as will be explained in the following. The characteristic parameters of each beamline are listed in Table 3.2.

3.2. ANKA and the standard XAS experiment

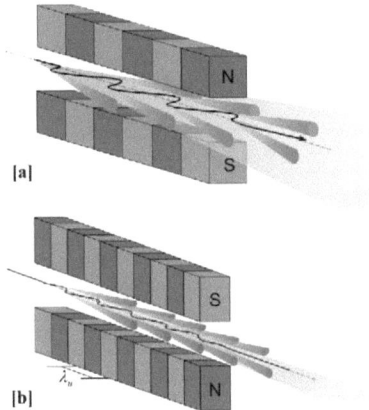
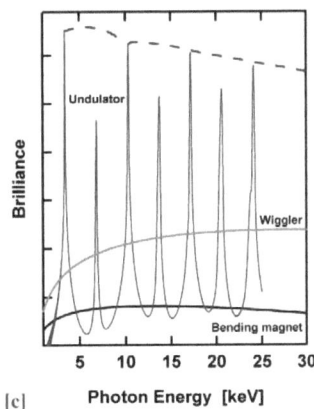

Figure 3.2.: Schematic description of a wiggler (a) and an undulator (b) as well as their spectral brilliance (in photons/s/mm^2/mrad^2/0.1%BW, where 0.1%BW denotes a bandwidth $10^{-3}\nu$ centered around the frequency $\nu = c/\lambda$, with c the speed of light and λ the photons wavelength) in comparison to a normal bending magnet (c) (figures taken from [35]).

3.2. ANKA and the standard XAS experiment

At ANKA, standard XAS measurements, in transmission and fluorescence mode, have been conducted at the INE beamline which belongs to the "Institut für nukleare Entsorgung" at the KIT. The corresponding setup shown in Fig. 3.4 (a) is, except for the final detector system, exemplary for all three beamlines that will be described. Not shown in this figure is the system of evacuated steel pipes, through which the synchrotron light, here emitted by the bending magnet, is directed. The synchrotron light is led through slits and mirrors for collimation and then directed into the Lemonnier [52] type double crystal X-ray monochromator (DCM), shown in Fig. 3.4 (c) – build by the Bonner synchrotron-group – where the desired X-ray energies are selected according to Bragg's law (see

Table 3.1.: Main parameters of the three synchrotrons.

	ANKA	DORIS III	ESRF
circumference [m]	110.4	289.2	844.4
Ring energy [GeV]	2.50	4.45	6.03
max. beam current [mA]	200	140	200
beam lifetime [h]	12	8	$30 - 80^a$
horizontal emittance [nm · rad]	41	450	4
dipole magnets bending radius [m]	5.559	12.181	23.366
dipole magnets field strength [T]	1.50	1.22	0.8
dipole magnets critical photon energy [keV]	6.0	16.04	20.5

a depending on filling pattern

3. Experiment

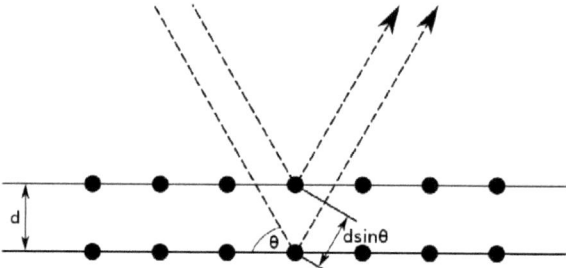

Figure 3.3.: "Bragg diffraction. Two beams with identical wavelength and phase approach a crystalline solid and are scattered off two different atoms within it. The lower beam traverses an extra length of $2 \cdot d \sin \theta$. Constructive interference occurs when this length is equal to an integer multiple of the wavelength of the radiation." [99]

Fig.3.3)

$$n\lambda = 2d \sin \theta \qquad (3.1)$$
$$\Leftrightarrow \quad E = hc/(2d \sin \theta) \cdot n, \qquad (3.2)$$

with $E = hc/\lambda$. Eq. (3.1) or Eq. (3.2) give the condition for constructive interference of two parallel waves, having a total path difference of $2d \sin \theta$ and that are scattered at a crystal. Here the integer n is the order of diffraction, λ the X-ray wavelength, d the distance between the lattice planes and θ the scattering angle. The interplanar lattice distance depends on the material and cut of the DCM crystals and is listed in Table 3.3, along with the corresponding energy range according to Eq. (3.2) and in due consideration of the allowed Bragg angles. For the Lemonnier type DCM the angle can be varied from 15° to 65°. An important point here is that at the INE beamline a Beryllium window is mounted in front of the DCM. The purpose of this is the separation of the storage ring's ultra-high vacuum from the subsequent devices, i.e. the DCM which is working at high vacuum of "only" 10^{-9} bar, allowing for crystal changes within manageable times. The DCM is followed then by a second focusing mirror and a second window to protect the high vacuum from the conditions at the experimental stage. The collimating and focusing mirrors, as well as the slit system (not shown), allow to further increase the intensity and focusing of the synchrotron radiation at the sample position.

The monochromatized X-ray radiation reaches the experimental stage (Fig. 3.4 (a+b)) and passes through a system of three ionisation chambers that allows for measuring XAS transmission spectra, according to the Lambert-Beer law in the form of Eq. (2.3), which is rewritten here:

$$\mu_x = -\ln(I/I_0) = \ln(I_0/I). \qquad (3.3)$$

The sample is positioned in between the first two ionisation chambers so that the transmission absorption coefficient μ_x (absorption μ times sample thickness x) is obtained, by measuring the X-ray intensities before (I_0) and behind it (I). Prior to the third ionisation chamber, a reference sample can be mounted that is measured simultaneously according to:

$$\mu_x^{ref} = -\ln(I^{ref}/I) = \ln(I/I^{ref}). \qquad (3.4)$$

Furthermore, a 5 pixel high purity germanium solid state fluorescence detector (Canberra Ultra-LEGe)

3.3. HASYLAB and the RIXS experiment

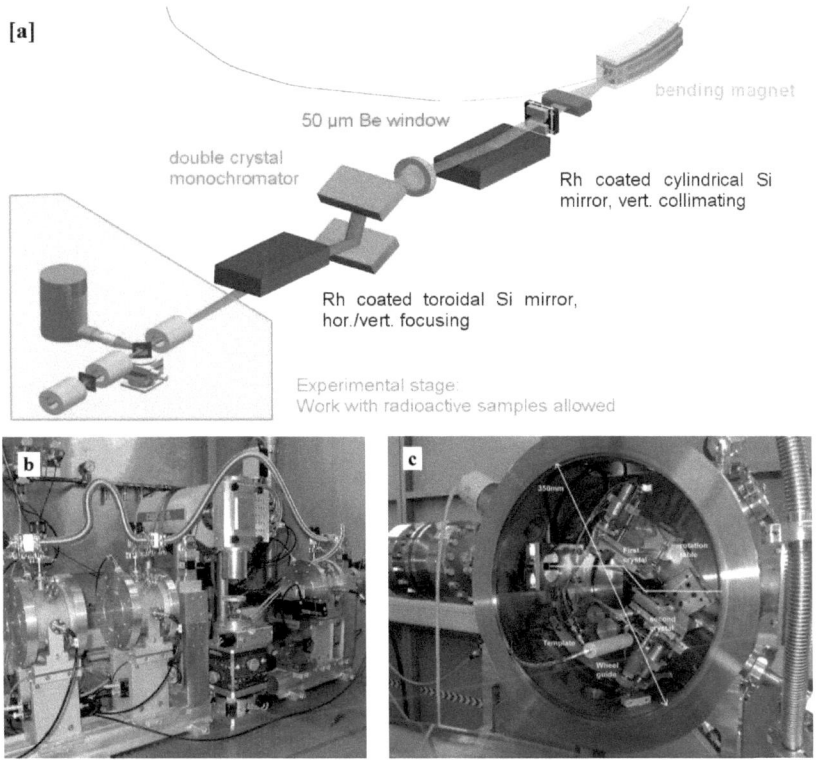

Figure 3.4.: Schematized setup of the INE beamline at ANKA (a), with photographs of the experimental stage (b) as well as of the double-crystal X-ray monochromator (c).

is stationed orthogonal to the beam direction and the sample, to provide (if sample is positioned appropriately) the detection of the emitted fluorescence radiation. This allows for fluorescence detected XAS measurements according to Eq. (2.5), i.e.:

$$\mu \propto I_f/I_0, \tag{3.5}$$

with I_0 as defined above and the fluorescence intensity I_f as measured by the detector.

3.3. HASYLAB and the RIXS experiment

At HASYLAB's W1 beamline resonant inelastic X-ray scattering (RIXS) measurements have been conducted. The beamline setup is shown in Fig. 3.5, and the main differences to the INE beamline are (1) the 32-pole wiggler as the source of synchrotron radiation that gives a one magnitude higher flux (see Table 3.2) and (2) the Johann spectrometer [41], installed inside a large vacuum vessel to minimize X-ray scattering by air, that allows to record emission spectra in dispersive geometry

3. Experiment

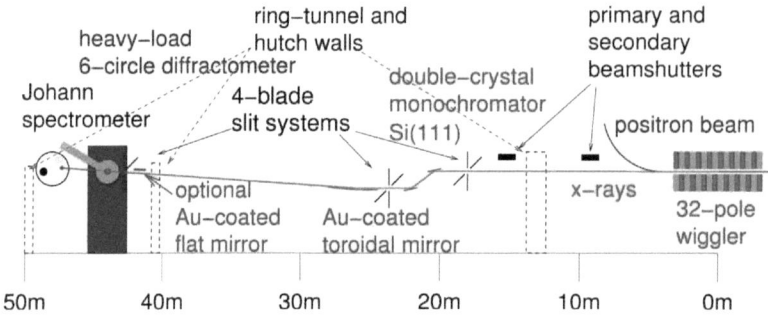

Figure 3.5.: Schematized setup of HASYLAB's W1 beamline at DORIS III (taken from [33]).

[96, 53]. In principle, also standard XAS measurements could be performed at W1, by just using the spectrometer to detect the total fluorescence and thereby neglecting its capability to highly resolve the fluorescence.

The dispersive geometry is briefly described with the help of Fig. 3.6: The incident X-rays, tuned by the DCM, are absorbed by the sample (S), and the subsequent emitted radiation (partly) reaches the analyzer crystal (AC). Here it undergoes Bragg diffraction and is focused towards the detector (D), as long as all three components S, AC and D are properly adjusted towards each other and positioned along the Rowland circle of radius $R = 0.5$ m. To further improve the observable energy range of the fluorescence radiation, the sample is moved somewhat inside the Rowland circle, so that its projected spot size on the circumference of the circle is increased. The Rowland geometry demands the spherically bent analyzer crystal to have a radius of curvature of $2R$, to assure diffraction with a fixed Bragg angle θ, independently of the diffraction position on the crystal (three positions are shown in Fig. 3.6). The dispersed fluorescence radiation is detected by the 1340 pixel × 1300 pixel (each pixel of size 20 μm × 20 μm) deep depletion CCD camera (EEV CCD36-40 with front-illuminated chip), with high resolution between 0.5 eV and 2 eV [34]. For the $K\beta_{1,3}$ line of cobalt, e.g., a Si(620) crystal is utilized (see Table 3.3 operation range) with a Bragg angle of 70.7°. The CCD camera as well as the energy dispersion of the crystal then allows the detection of an emission energy range of 75 eV around the emission peak, with energy resolution of about 1 eV, determinable from the full width half maximum

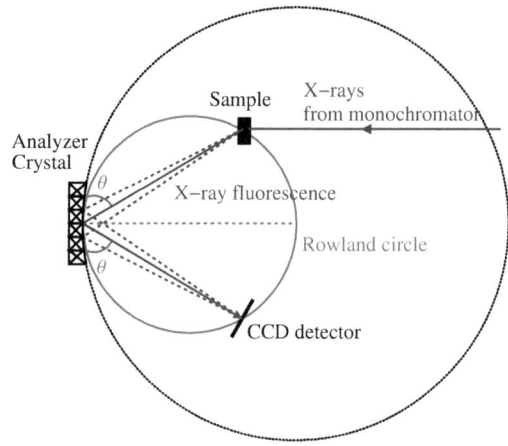

Figure 3.6.: Schematized Johann spectrometer: Detection of fluorescent X-ray radiation in Rowland geometry (circle of radius R) by a spherically bent crystal analyzer with radius of curvature $2R$.

3.4. ESRF: RXES and HRFD-XAS experiments

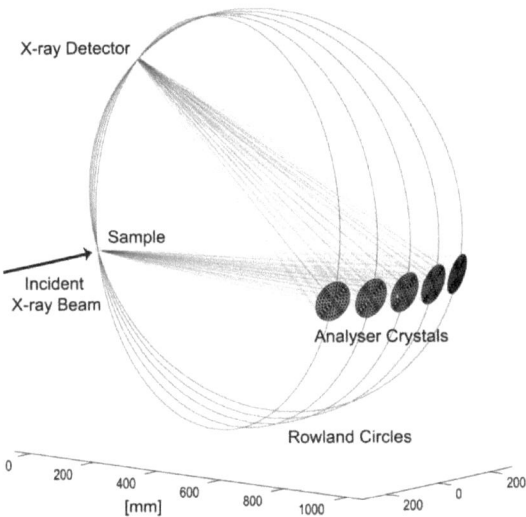

Figure 3.7.: Rowland geometry as utilized at the ID26 beamline at the ESRF. The red lines denote the fluorescent radiation which is emitted from the sample and then Bragg diffracted by five analyzer crystals and focused onto the detector (taken from [28]).

(FWHM) of the elastic peak

A standard measurement at W1, gives a three dimensional RIXS plane (as was shown in Fig. 2.11), where each intensity point depends on excitation and emission energy, and whereof two dimensional spectra can be extracted, as was described already at the end of section 2.3.

3.4. ESRF: RXES and HRFD-XAS experiments

At the ESRF the ID26 beamline (see Table 3.2 for details) was visited. It is an insertion device source consisting of three mechanically independent undulators (one 40-mm period and two 35-mm period), yielding two orders of magnitude higher flux compared to HASYLAB's W1. The beamline setup is similar to the W1 beamline, in that the fluorescent radiation from the sample is Bragg diffracted by an analyzer crystal in Rowland geometry as can be seen in Fig. 3.7, i.e. it is also a Johann spectrometer [41]. However, here by contrast five analyzer crystals are utilized, allowing to collect a larger amount of the fluorescence radiation which is focused onto a single avalanche photo diode (APD) that records one certain fluorescent energy only, in contrast to the CCD detector in dispersive geometry.

3. Experiment

Table 3.2.: Main parameters of the visited beamlines hosted at the synchrotrons and described in Table 3.1. For more information about the crystals used, see Table 3.3.

		INE (ANKA)	W1 (DORIS III)	ID26 (ESRF)
source	type	bending magnet	wiggler	3 undulators
	critical energy [keV]	6.0	8.1	9.8
	energy range [keV]	2.1 – 25	4.0 – 11.5	2.5 – 30
beam	flux at sample [ph./s/mm^2]	2×10^{10} [a]	2×10^{11} [b]	$> 10^{13}$ [c]
	size at sample [mm×mm]	0.5 × 0.5	4.0 × 1.6	0.25 × 0.15
XAS	monochromator type	Lemonnier DCM	DCM	DCM
	energy resolution [$\Delta E/E$]	2×10^{-4} [a]	2×10^{-4} [b]	1.4×10^{-4} [c]
	crystals used	Si(111), Ge(422)	Si(111)	Si(111)
XES	spectrometer type	5 pixel fluo. det.	Johann spectrometer	Johann spec.
	energy resolution [eV]	~ 10	~ 1	~ 1
	crystals used	Ge solid state	Si(111), Si(531), Si(620)	Ge(444)

[a] at 18 keV using Ge(422), [b] at 9 keV using Si(111), [c] at 9 keV using Si(111)

Table 3.3.: Crystals used by DCMs and spectrometers in this work. a is the lattice constant and $d = a/\sqrt{h^2 + k^2 + l^2}$ the interplanar distance with respect to the cut defined by the miller indices h, k, l. The energy ranges possible with these crystals are calculated based on Eq. (3.2), with $n = 1$ and for the DCM with $\theta = 15° - 65°$ (E_{ex}) and for the Johann spectrometer with $\theta = 60° - 86°$ (E_{em}).

material	a [Å]	cut	d [Å]	range E_{ex} [eV]	range E_{em} [eV]
Si	5.43095	111	3.1356	1344 – 7639	–
		311	1.6375	4177 – 14627	–
		333	1.0452	–	5946 – 6849
		531	0.9180	–	6769 – 7685
		620	0.8587	–	7237 – 8336
Ge	5.64613	422	1.1525	5935 – 20783	–
		444	0.8149	–	6779 – 7798

4. Synthesis of Nanoparticles

As one of a particle's dimensions gets into the nanoscale, i.e. between 1 and 100 nm, it is called a "nanoparticle". The emphasis is onto "one dimension", which implies that the particle could be, e.g., much longer and/or wider than 100 nm, but very thin in one or two dimension(s) resulting in a "nanolayer" or "nanowire", respectively. If one of the dimension of the nanoparticles is even between 1 and 10 nm and exhibit a narrow size distribution, which means that the average size must vary only about 20 %, one talks of "nanoclusters". If the sizes are distributed only about 10 % around the average value, the particles are termed "monodisperse". Nanoparticles or -clusters, which are not amorphous, are in most cases consisting of several crystalline regions. If even the complete particle is single-crystalline, it is declared a "nanocrystal" [103].

Anyway, although the terms above are the "official" definitions, the word nanoparticle usually is used in the literature to describe particles with all three dimensions in the nanoscale, whereas nothing is stated about the shape, which could be spherical, cubical, polyhedral, or anything else. If one or two dimensions are considerably larger than 100 nm, in the literature one talks of nanowires or nanolayers, as aforementioned. In this work, the term "nanoparticle" will be used for all nanoscaled (mainly spherical) particles, although they are (at most) monodispersed and between 1 and 10 nm in diameter (and which is not uncommon either).

For the creation of metallic nanoparticles two general approaches can be distinguished, "top-down" and "bottom-up", which are schematized in Fig. 4.1. In the top-down approach nanoparticles are build by the structural decomposition of a solid body, i.e. from large entities to nanosized ones, which involves ball milling or the powerful technique of lithography. Currently, the top-down methods are superior with respect to the possibility of building electronic circuitries. On the other hand, in bottom-up approaches nanoparticles are created by nucleation of molecules (or atoms, if possible), i.e. from the smallest building blocks to nanosized ones and thus, in principle, with atomic precision (see e.g. [16]). Here one can further divide into gas-phase and wet-chemical (or sol-gel) techniques, referring to the aggregate state of the initial material, gaseous or liquid.

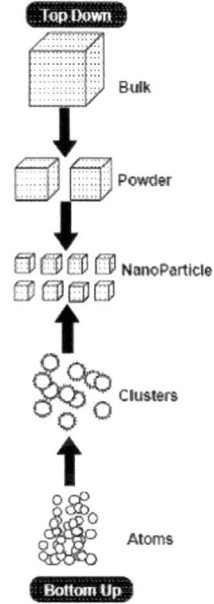

Figure 4.1.: Two general approaches in nanoparticle synthesis.

The fundamental process of the wet-chemical synthesis is the creation of smallest fragments, from a precursor material, in a solution, which then cluster to form nanoparticles. To control the overall chemical process and thereby the nanoparticles' size and shape, a so-called stabilizing agent (or reactant/surfactant) is used, which has the following important functions (compare [37]): (1) prevents the particles from agglomeration so that stable colloids can be formed, (2) controls the speed and

4. Synthesis of Nanoparticles

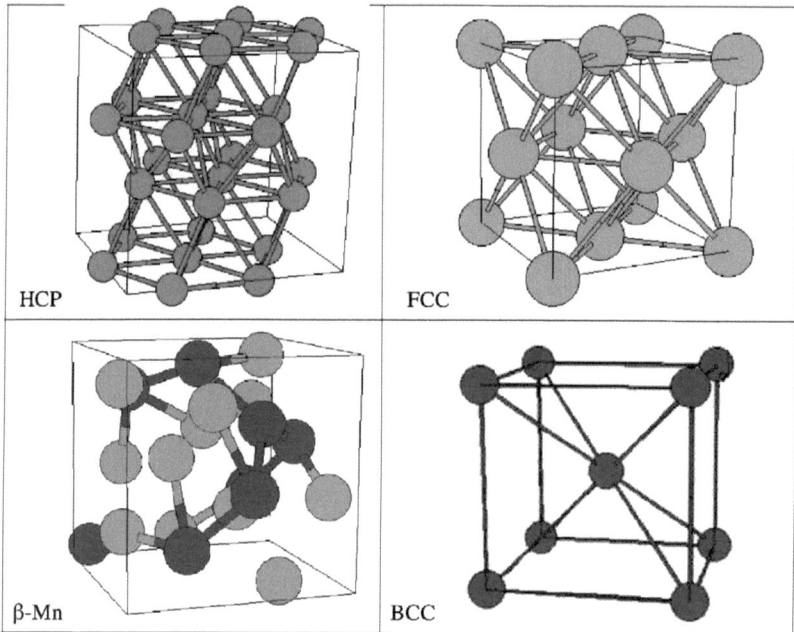

Figure 4.2.: Possible cobalt phases: Co-hcp, Co-fcc, Co-ε (β-Manganese), and Co-bcc.

duration of nucleation, thereby determining the size of the particles, (3) can selectively be attached to different crystal surfaces during the growth to control the particle shape and (4) determines the particle's polarity. An important point is that the surfactants can contain functional groups which interact weakly or strongly with the nanoparticles, thereby preserving the electronic properties of the "naked" nanoparticles, or altering them by charge redistribution (on the nanoparticles surface) so that new (wanted) properties can arise, like self-assembling [92] (compare [114]).

One might argue now that it should be possible to tailor nanoparticles with respect to their designated field of application, just by controlling the synthesis route by means of surfactants and other parameters, like heating temperatures, duration of time intervals etc. However, in numerous experiments it was revealed that it is not possible to separately control the various properties of the final product. In fact, it has become evident that there is a strong interdependence of those properties and that it is quite difficult to change one property in a systematic manner without influencing the others [37, 114].

4.1. Cobalt nanoparticles

Wet chemical techniques offer the greatest flexibility for fabricating nanoparticles, and as the surfactants also have a pronounced effect on the crystal structure of the nanoparticles, it could be shown for cobalt that (at least) three different crystallographic forms can be synthesized with combinations of different surfactants [86]: hexagonal-close-packed (hcp), face-centered-cubic (fcc) and β-Manganese

4.1. Cobalt nanoparticles

(ϵ). The latter phase was first reported in [87] and later in [22]. Additionally, meta-stable Co nanoparticles exhibiting the body-centered-cubic phase (bcc) were discovered [76] and investigated thoroughly [31]. All four Co phases are shown in Fig. 4.2.

4.1.1. Co nanoparticles for high-resolution X-ray measurements

The synthetic methodology for the preparation and anti-corrosive stabilisation of zerovalent cobalt nanoparticles was recently reviewed in detail by Bönnemann et al. [7]. Briefly, air stable Co nanoparticles are obtained by thermolysis of $Co_2(CO)_8$ in the presence of the stabilizing agent $Al(C_2H_5)_3$ (atomic ratio Co:Al = 10:1) and with monodispersed size distribution of 6 nm, as was determined by transmission electron microscopy (TEM) (see Fig 4.3). The single synthesis steps have been: (1) The solid $Co_2(CO)_8$ precursor is given into a toluene bath, by addition of the stabilizing agent $Al(C_2H_5)_3$ and is mechanically stirred at room temperature, until all $Co_2(CO)_8$ is dissolved. (2) The mixture is heated to 110° and later up to 150° and kept stirring for strictly defined time intervals under toluene reflux, to allow the continuous formation of the Co nanoparticles. (3) After cooling down to room temperature, a slow stream of (a defined amount of) diluted oxygen (4.5 % O_2 in N_2) is introduced into the suspension of nanoparticles, with the aim to form a thin protective shell. (4) When the "smooth oxidation" process is finished, stirring is stopped to allow the nanoparticles to settle down. At last, the supernatant will be decanted and the Co nanoparticles are washed several times with toluene, to get rid of all remnants from the precursor and stabilizing agent.

To avoid possible oxygen contamination, the whole synthesis is performed under protection gas (pure Argon) and the final nanoparticles are stored in wet form in toluene, until the beginning of the X-ray measurements. For the measurements they are dried by employing a vacuum pump, and the final nano-powder is prepared (sealed in Kapton) under Ar/N_2 atmosphere in a glove-box.

To have Co nanoparticles at hand with different ratios of metallic core to protective shell, step (3) of the synthesis have been varied with respect to the speed of the reduced oxygen stream and its total amount. The result have been three types of Co nanoparticles that are identically (with respect to the other synthesis steps), but with a thin protective shell ("smooth oxidation"), an intermediate shell ("controlled oxidation") and a thick shell ("rough oxidation").

For the measurements at HASYLAB (section 6.2.2) only nanoparticles with the thickest shell were utilized, as they offer the most uniform ratio of the two Co compounds (about 50 : 50 as will be shown in section 6.2.2). Furthermore, equally synthesized Co nanoparticles have been exposed to air for about one month, to impose complete oxidation. Hereby, a reference for the shell of the Co nanoparticles is provided. For the final investigations at the ESRF (section 6.3) all three types of nanoparticles – smoothly, controlled and roughly oxidized – were available for measurements.

Noteworthy is that $Al(C_2H_5)_3$ is chosen solely as stabilizing agent/surfactant to have the procedure and the elements involved as simple as possible. The protection from further agglomeration and oxidation is assured due to the smooth oxidation (protection gas is utilized nonetheless as a precaution), so that no additional coating is necessary and the number of metal sites present are minimized. As was proposed in [77], the smooth oxidation process leads to the formation of a thin protective shell layer of probably CoO and/or $CoCO_3$, around a metallic (dominantly fcc) core. The crystallographic structure of the current Co nanoparticles is expected to be different from that in [77] though, as they have applied Octyl ($Al(C_8H_{17})_3$) in contrast to Ethyl in this work.

4. Synthesis of Nanoparticles

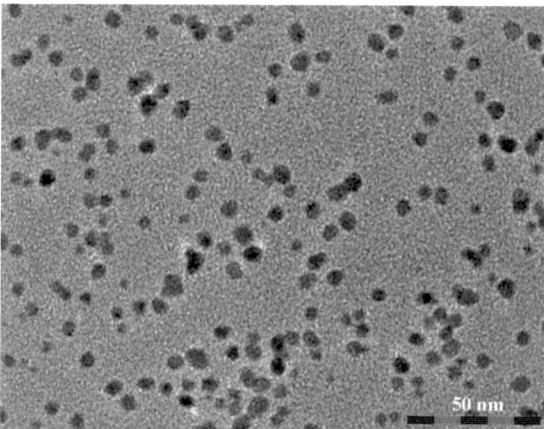

Figure 4.3.: TEM picture of the Co nanoparticles.

4.2. Cobalt-Platinum nanoparticles

4.2.1. Co_3Pt/C nanoparticles as catalysts

The synthesis of the Co_3Co/C nanoparticles that are utilized as electrode material in a proton exchange membrane fuel cell (PEMFC), is thoroughly described in Ref. [81]. The synthesis starts as described in section 4.1.1 steps (1) and (2), except that Al-Octyl is used instead of Al-Ethyl. The final Co nanoparticle powder then is dispersed in toluene again, but with addition of Carbon Black (3 times the quantity of the initial cobalt-carbonyl) and is sonicated for nearly one day. The resulting Co/C powder eventually is soaked in a solution of $Me_2Pt\,COD$ (prepared according to [97]) for some time, dried afterwards and treated by hydrogen at 60°C in the "conditioning process" [10], leading to separated Co and Pt nanoparticles attached to C [81]. The addition of Carbon Black has two functions: Once to enhance the conductivity when applied in a fuel cell and further on to prevent unrestricted agglomeration, as it serves as substrate where the Co-Pt nanoparticles get attached to. However, it does not protect the nanoparticles from contamination when exposed to air, so that all further treatment must be performed air-protected.

The final Co_3Pt/C nanoparticles catalyst was heat-treated at temperatures between 350 °C and 1000 °C. High-resolution TEM images were taken for all the resulting nanoparticle states – the 'as prepared' and the 800 °C tempered particles are shown in Fig. 4.4 – whereof sizes were determined by H. Schulenburg *et al.* [81].

4.2. Cobalt-Platinum nanoparticles

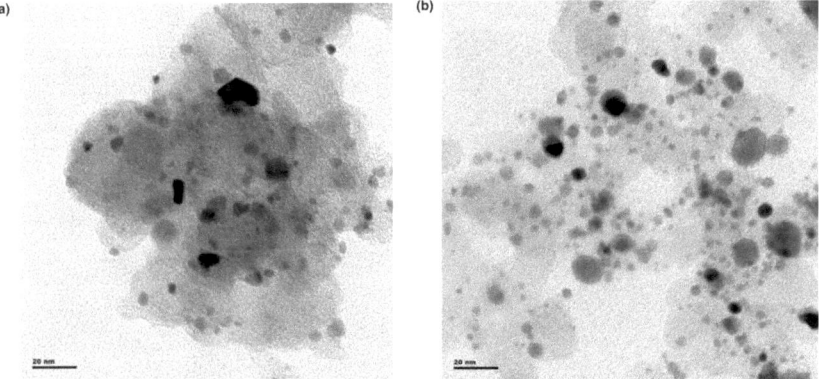

Figure 4.4.: HRTEM images of Co$_3$Pt/C nanoparticles in the states 'as prepared' (a) and tempered to 800 °C (b) (pictures taken from [81]).

5. Co-Pt Nanoparticles as Catalysts in Fuel Cells

"A fuel cell is an electrochemical cell that converts chemical energy from a fuel into electric energy." [101] Fuel cells are already used as stationary energy generators in hospitals, hotels, schools, etc. and portable versions are planned for laptops, cellular phones and other mobile devices [1]. The main purpose, however, is the usage in transportation, to replace the classic combustion cycle, where some fossil fuel is burnt to run an engine, and a lot of environmentally hazardous waste is emitted as by-product. Fuel cells that consume (the most common fuel) hydrogen (H_2) and air, leading to the exhaust product of just water, are mainly investigated thus. As pointed out recently in Nature [91], there are four major challenges that go along with the practicable establishment of fuel cells:

1. Hydrogen production: Most common still is the extraction of hydrogen by steam reforming of natural gas, whereby carbon dioxide (CO_2) is produced as well, so that nothing is won with respect to the classic combustion engine. The balance gets even worse when the power source to perform the H_2 production is taken into account, which in general produces (a lot of) CO_2 too. Thus, the consequent development of carbon-free power plants, by which a carbon-free extraction of hydrogen, e.g., from water via electrolysis, is maintained at an industrial-scale, is mandatory.

2. Hydrogen storage: The previously preferred liquid hydrogen that constitutes the most dense state of H_2 was already ruled out due to the required cooling down to $-253\,°C$. Instead, carbon-fibre tanks were developed that allow to store hydrogen at a pressure of about 680 atmospheres, allowing a car to drive several hundreds of kilometres.

3. Infrastructure: "Fuel-cell vehicles will never sell in a big way until there is a viable network of service stations to fuel them." [91] This is an important point that, however, is strongly determined by politics and therefore will not further be elaborated.

4. The fuel cell: The general design of a fuel cell is shown in Fig. 5.1 and, simplified, just is comprised of an anode, electrolyte and a cathode, packed together. The fuel H_2 oxidizes at the anode and the H^+ ions pass through the electrolyte (a proton or polymer exchange membrane, PEM), while the electrons are blocked by it and thus travel through a wire generating an electric current. At the cathode the O_2 of the air is reduced (Oxygen reduction reaction, ORR), then also blocked by the membrane, to finally combine with the electrons and ions to become water which is withdrawn. Such a fuel cell is called PEM fuel cell (PEMFC). An important point for all fuel cells is the cost per unit of energy that is produced, which is strongly correlated to the electrodes' material, which for the PEMFC (and most other fuel cells too) is platinum. Three objectives go along with platinum:
    a) As Pt is on of rarest elements and consequently very expensive, a major aim is to reduce its amount necessary for the electrodes or even (partially) replace it by non-precious metals.

5. Co-Pt Nanoparticles as Catalysts in Fuel Cells

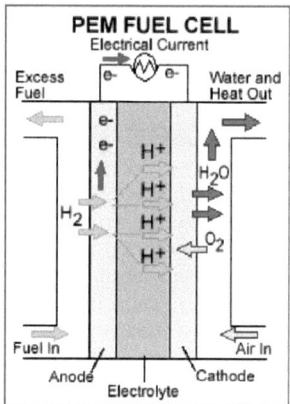

Figure 5.1.: Diagram of a proton exchange membrane fuel cell (PEMFC) (picture taken from [105]).

b) Pt is poisoned by carbon monoxide (CO) that blocks its catalytic active sites as it is strongly adsorbed. CO is generally present in H_2 gas produced from natural gas, which still is the only source used for mass-production. Here a range of attempts have been made, e.g., by alloying Pt with 3d or 4d transition metals. Two main origins are mentioned in the literature that decrease the poisoning by CO, which both result from a weakened surface Pt–O bond strength: Either the adsorbed CO is oxidized via an O adatom (to build CO_2 that does not contaminate Pt), or CO is dissociated at the Pt surface. The former effect was found for, e.g., Ru–Pt core–shell nanoparticles [2] and even both effects for Pt–Co alloys, whereby the strength of each effect was dependent on the Pt–Co composition (which was varied) of the surfaces [38].

c) Lastly, it is always essential that while realizing a) and/or b), the overall catalytic activity must increase simultaneously – which is in general always the case for b) (albeit 100 % pure H_2 is used) – as the fuel cell still lacks efficiency, mainly due to the slow kinetics of the ORR [62].

The focus of this work is on point 4. c), i.e. increase of the catalytic activity and investigation of the reason for the increase. Furthermore, as this will be realized via Co-Pt nanoparticles (attached to a carbon (C) support), the reduction of the amount of Pt is implicit since the particles are nanoscaled with Pt fraction of only about 25 %. The Co_3Pt/C nanoparticle catalyst was developed by H. Bönnemann and G. Khelashvili [12] (see also section 4.2.1) and described in dependence on heat treatment, among others, via high-angle annular dark-field scanning transmission electron microscopy (HAADF-STEM) combined with energy-dispersive X-ray (EDX) spectroscopy and further on by X-ray diffraction (XRD) by H. Schulenburg et al. [81]. By rotating disk electrode (RDE) measurements, a maximum mass activity 2.4 times higher than for commercial $Pt_{2.5}Co/C$ was found, for samples tempered at 800 °C. No correlation between activity and nanoparticle size or surface area was observed. Instead, the highest activity was related to the presence of the face-centred-tetrahedral (PtCo-fct) intermetallic compound $L1_0$, with lattice constants $a = 2.691$ Å and $c = 3.684$ Å, as well as to an ordered face-centred-cubic 50 : 50 alloy phase of Pt and Co (PtCo-fcc), with lattice constant $a = 3.744$ Å. Both phases PtCo-fct and fcc were found to be dominant from 700 °C to 1000 °C and from 800 °C to 1000 °C, respectively, which were found to be the regions of highest catalytic activity.

5.1. XANES measurements of Co₃Pt/C nanoparticles

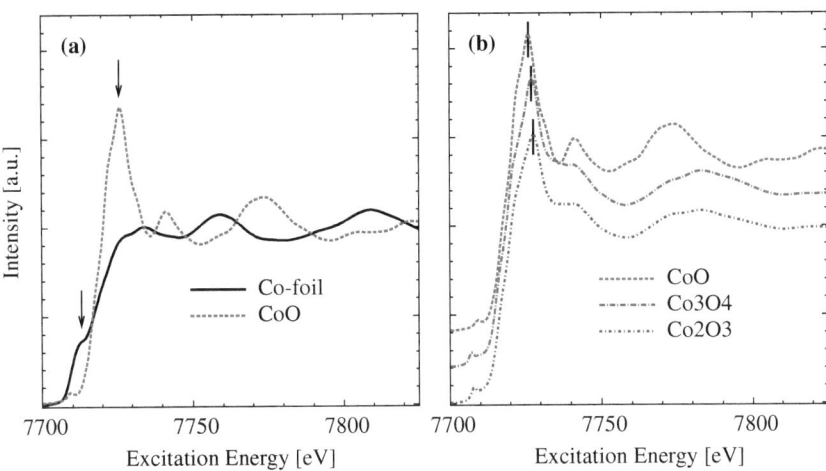

Figure 5.2.: Normalized Co K-edge XANES spectra of references Co-foil and CoO (a) (edge shoulder and whiteline marked by arrows), as well as CoO, Co_3O_4 and Co_2O_3 (b) (with whiteline positions marked by vertical lines).

As the measurements of H. Schulenburg et al. [81] unveiled no definite origin for the strong increase of the catalytic activity of the Co_3Pt/C nanoparticles at 800 °C, XAS measurements have been conducted to complement the study. XAS will give insight into the electronic (valence) structure and the local atomic environment, which was not inspected yet, for crystalline and amorphous components of the nanocatalyst.

5.1. XANES measurements of Co₃Pt/C nanoparticles

The X-ray absorption spectroscopy (XAS) measurements were performed at the INE beamline at ANKA (see section 3.2) in transmission mode. A Lemonnier type [52] double crystal monochromator (DCM), equipped with a pair of Si(111) crystals for the cobalt K-edge at 7709 eV and Ge(422) crystals for the platinum L3-edge at 11564 eV, was employed with energy resolution $\Delta E/E = 2 \times 10^{-4}$. The DCM was tuned from −150 eV to about +850 eV (corresponds to $k \simeq 15\,\text{Å}^{-1}$) relative to the edges, to record XANES and EXAFS. For the XANES region from −30 to +30 eV (relative to the edge) the stepsize have been set to 0.5 eV, with collection time $t = 1$ s for each step, and for the adjoining EXAFS region the stepsize was defined as $\Delta k = 0.03\,\text{Å}^{-1}$, with collection time increasing with increasing k value to account for the oscillations that become successively weaker at high energies (or k values). Co_3Pt/C nanoparticles were investigated 'as prepared' (Co3PtC-asprep), completely without Pt (CoC-asprep), and in the 350 °C (Co3PtC-350) as well as 800 °C tempered state (Co3PtC-800). Further on, a commercial catalyst Pt_3Co/C from TKK (Pt3CoC-TKK), as well as common Pt references, metallic Pt (Pt-foil) and Pt(IV)O (PtO₂) in powder form, and also Co references, Co-foil and powdery Co-oxides Co(II)O (CoO), Co(II,III)O (Co_3O_4) and Co(III)O (Co_2O_3), have been measured. All nanosized samples have been handled air-protected during all processes, to avoid oxygen contamination.

5. Co-Pt Nanoparticles as Catalysts in Fuel Cells

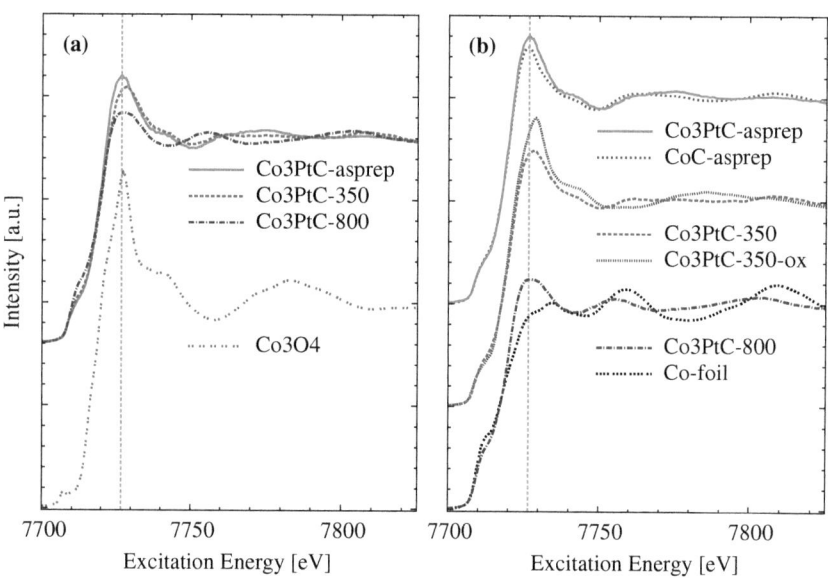

Figure 5.3.: Normalized Co K-edge XANES spectra of the Co_3Pt/C nanoparticle catalyst 'as prepared' as well as tempered to 350 °C and 800 °C in comparison to references.

At the Co K-edge the formal oxidation state and thus the metallicity of the Co nanoparticles can be determined by well known effects, as will be demonstrated with the help of the XANES spectra in Fig. 5.2. Here and in the following, all XANES spectra are normalized as described in the appendix C. In Fig. 5.2 (a) metallic Co (valency 0) is compared to CoO (valency +2) and the following changes can be observed when going from valency 0 to +2: The edge shoulder at about 7713 eV moves to higher energies, a pre-edge feature arises at about 7709 eV and a strong whiteline at about 7726 eV arises. Further on, the shape resonances at higher energies are completely changing which, however, is not anymore due to changes in the electron structure but is a result of a geometrical transition. In Fig. 5.2 (b) the stable Co-oxides are compared. They show a successive energy shift of the whiteline with increase of formal oxidation state (+2, +2/+3 and +3). To understand these effects one has to take a deeper look into the electron structure: As soon as Co is part of a compound, like in CoO, the metal valence bands (3d, 4p) are shifted to higher energies due to the attraction by the O ligands that are included into the Co lattice, and the energy difference towards 1s is increased [6]. Consequently, the edge-shoulder resulting from 1s→4p transitions is shifted to higher energies, merging eventually into the whiteline. The whiteline is much stronger for CoO than for Co-foil due to new electron states, resulting from hybridization of Co-4p and O-2s and 2p orbitals, and its energetic position shifts with the valency by about 2 eV from +2 (CoO) to +3 (Co_2O_3), due to an increased amount of O in the Co lattice (Co_3O_4 formally is just a mixture of CoO and Co_2O_3). Lastly, due to the increasing amount of Oxygen 2p orbitals overlapping with the metal 3d states, the transition 1s→3d becomes partly dipole-allowed, leading to a small but sharp pre-edge feature (compare [110] for assignments of the features). Thus, by comparing the energy position and intensity of the edge shoulder and whiteline of Co-compounds with metallic Co and Co-oxides, the metallicity and valency can be determined. It is

5.1. XANES measurements of Co₃Pt/C nanoparticles

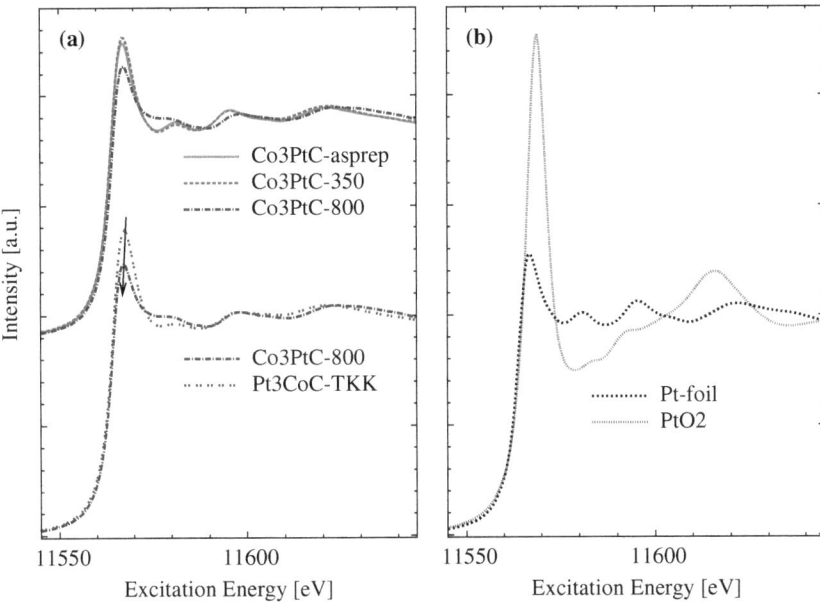

Figure 5.4.: (a) Normalized Pt L3-edge XANES spectra of the Co₃Pt/C nanoparticle catalyst 'as prepared' as well as tempered to 350 °C and 800 °C (upper curves) and the latter compared to commercial Pt₃Co/C catalyst from TKK (lower curves). (b) Normalized Pt L3-edge XANES spectra of references Pt-foil and PtO₂.

important to keep in mind, however, that XANES spectra are reflecting not only the electron, but also the crystal structure which, if altered, can also modify the features sensitive to the electron density of states (pre-edge etc.) [29].

The normalized Co K-edge XANES spectra of the Co₃Pt/C nanoparticle catalyst in three states are shown in Fig. 5.3 (a) and (b). In (a) on can see that significant changes occur upon tempering the catalyst to 800 °C. From (b) it is clear that Co₃Pt/C-asprep is almost identical to Co/C-asprep, indicating that the low edge and strong whiteline are due to bonds with carbon only - and not due to (unwanted) oxidation, which de facto is not possible anyway, as the nanoparticles were handled strictly air-protected all the time. To further check this, both samples had been remeasured after being exposed to air for some days (not shown), and clear changes as a result of oxidation were visible in the XANES spectra (the whiteline increased even more, but peaked at different energy) and in the EXAFS spectra (the visible shell structure resembled a typical Co-oxygen in contrast to the samples not exposed to air). Coming back to Fig. 5.3 (b), a formal oxidation state similar to Co₃O₄ can be declared for Co₃Pt/C-asprep, i.e. a valency between +2 and +3 (indicated by vertical grey dashed line). In (b) one can also see that the nano-catalyst, when oxidized (Co₃Pt/C-350-ox), adopts an even higher valency, at least +3 (when compared to Co₂O₃). Finally, Co₃Pt/C-800 obviously is in a metallic state as its edge shoulder at about 7713 eV resembles that of Co-foil, however, with significantly different crystallographic structure, recognizable in the different whiteline and shape resonances at

5. Co-Pt Nanoparticles as Catalysts in Fuel Cells

higher energies.

In Fig. 5.4 (a) normalized Pt L3-edge XANES spectra of the nanoparticle catalyst in the three states ('as prepared', tempered to 350 °C and 800 °C) are shown and in (b) the references Pt-foil and PtO_2. The increase in intensity from small to large energies, peaking at about 11567 eV in the whiteline, is due to electron transitions from $2p_{3/2}$ to $5d_{5/2}$. Thus, the more vacant the 5d valence band, the stronger the whiteline, which is most noticeable when comparing Pt-foil with PtO_2, with electron configurations [Xe]-$4f^{14}$-$5d^9$-$6s^1$ and -$5d^6$-$6s^0$, respectively. Significant electronic and crystal changes are visible in Co_3Pt/C-800, whereas from room temperature ('as prepared') to 350 °C no modifications are visible in Pt-L3 XANES. It is obvious that the, with respect to the ORR, most active nanoparticles Co_3Pt/C-800 exhibit the lowest whiteline, which is also valid when comparing to a commercial Pt_3Co/C catalyst (from TKK), shown in the lower part of Fig. 5.4 (a). This lowering of the whiteline intensity is accompanied by a slight shift to lower energies, indicated by an arrow. It can be attributed to an increased occupancy of the 5d electron levels, as the result of a lowered 5d band center, which is a known ORR activity increasing effect [111]. The lowering of the 5d band center is related to a "weakening of the metal–oxygen bond strength, which is induced both by a shortened Pt–Pt bond distance and by the ligand effect" [17]. The former is a result of strained surface layers due to the differing lattice constant of the second metal (Co in this case), and the latter comprises the interactions of the ligands (as part of a coating or substrate) with the metal(s), leading to changes in the chemical properties of the surface [45]. The weakened metal–oxygen bond strength allows for a faster ORR, as the adsorbed O_2 will be less tight bound, but nonetheless dissociated. Of course, a limit exists where the bonding gets too weak, thereby hindering the dissociation.

5.2. EXAFS measurements of Co_3Pt/C nanoparticles

The Co_3Pt/C nanoparticle catalysts' EXAFS spectra are shown in Fig. 5.5 in k- as well as R-space and at both edges, Pt-L3 and Co-K. Here, k is the photo-electrons' wave vector according to Eq. (2.6), and R-space spectra are obtained by a Fourier transformation (Eq. D.1 in appendix D) of the k^2-weighted k-space spectra in the range $k = (3 - 12)$ Å^{-1}. Here, the lower limit corresponds to a position a few eV above the whiteline as well as the first shape resonance of all samples, to assure that no bound states are involved since the EXAFS analysis relies on the simulation of continuum states. The upper limit just gives the borderline where the data quality gets too worse. More details about EXAFS and the pre-processing of EXAFS data can be found in appendix D. Co_3Pt/C tempered to 800 °C shows the most drastic changes as was recognized via XANES already. However, in contrast to XANES, significant differences are already visible at 350 °C compared to the 'as prepared' state.

In order to correlate these spectral changes to the atomic structure of the Co_3Pt/C nanoparticles, the EXAFS spectra have been analysed with the help of the Artemis software [69]. Artemis allows for fitting experimental EXAFS spectra path-by-path in k- or R-space according to Eq. (2.33), whereby the paths are simulated by FEFF [75]. R-space was chosen as fitting space, as only a section from 0.8 to 6.0 Å was fitted (coordination shells 1 – 5), i.e. not the complete k-space oscillations were utilized (see appendix D for further details).

The three states of the Co_3Pt/C nanoparticles have each been fitted on both edges simultaneously, since all hetero-metallic paths – i.e. electron scattering at Co-atoms, when Pt is the absorbing atom and vice versa – are visible and equivalent at both edges. For Co_3Pt/C-800 the fit is shown in Fig. 5.6 and the detailed results are given in Tables 5.1 and 5.2, along with those from the simultaneous fits of Co_3Pt/C-350, Co_3Pt/C-asprep, as well as fits of Pt-foil and Co-foil for reference. In Fig. 5.6 the main contributions for each coordination shell are shown as well. Pt–Pt, for example, denotes

5.2. EXAFS measurements of Co_3Pt/C nanoparticles

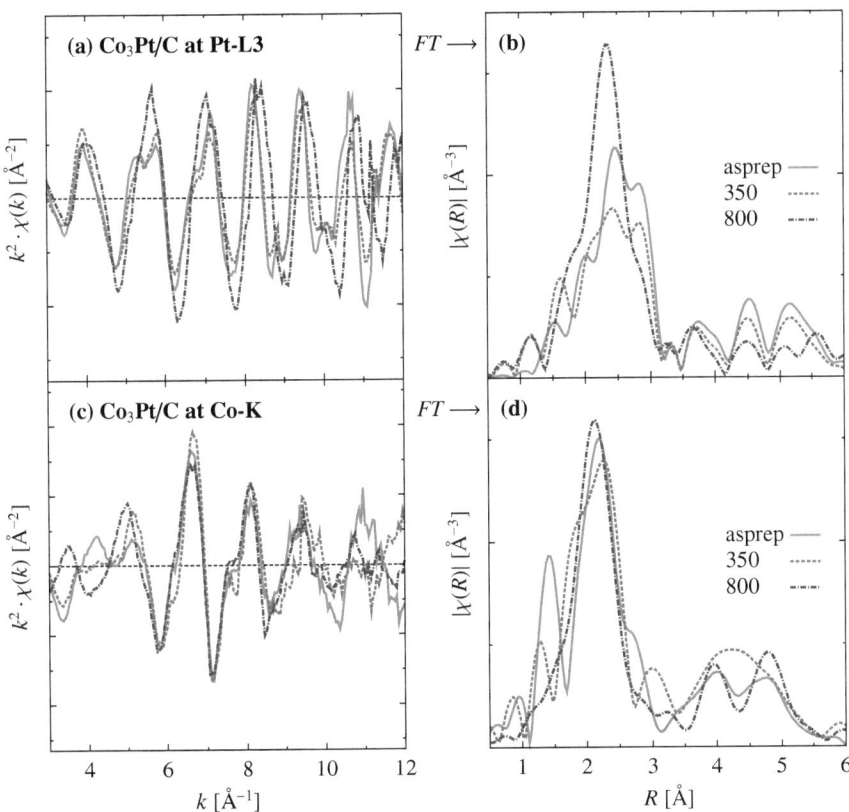

Figure 5.5.: EXAFS spectra of the Co_3Pt/C nanoparticle catalyst in three states at both edges. k^2-weighted k-space spectra on the left and respective Fourier transformed ($FT\longrightarrow$) R-space spectra on the right.

single scattering of the photo-electron, ejected from the absorbing Pt atom, scattered at another Pt atom (belonging to the respective shell) and travelling back to the absorber. For the fourth shell the single-scattering gives only a marginal contribution, instead the collinear double (and triple) scattering paths are dominant. For the three states of the nanoparticles – 'as prepared', 350 °C and 800 °C – the same fitting model has been applied: a 50 : 50 fcc Pt-Co alloy with parameters (P), energy shift ΔE_0 for each edge ($\rightarrow$ 2 P), Debye-Waller (DW) factors σ_i^2 for each shell ($\rightarrow$ 5 P), coordination-number reduction factor δN (that allow to determine total coordination numbers N_i) for Pt/Co–C as well as all Pt–Pt and Pt–Co and Co–Co shells, respectively, ($\rightarrow$ 5 P) and distances R_i for each shell at both edges, whereby the Co–Pt distances were kept equal at both edges ($\rightarrow$ 17 P). For the simultaneous fits this gives a total of 30 parameters versus about 59 independent points for two EXAFS spectra (calculated according to Eq. (D.3) in appendix D). The amplitude reduction factor S_0^2 was determined once by fitting the Co-fcc phase to Co-foil ($S_0^2 = 0.76$) and Pt-fcc to Pt-foil ($S_0^2 = 0.83$) and then was

5. Co-Pt Nanoparticles as Catalysts in Fuel Cells

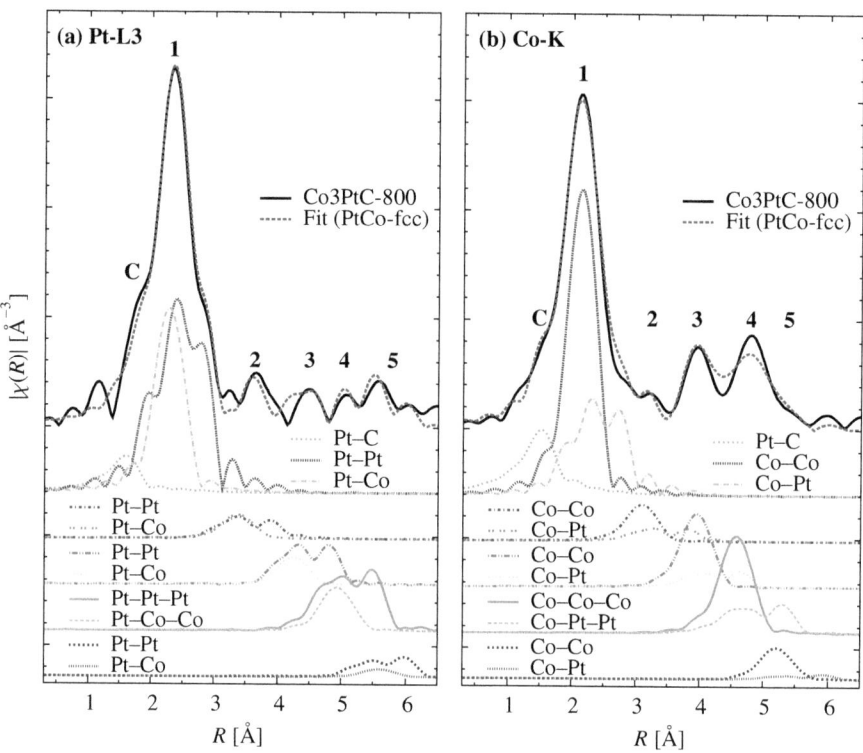

Figure 5.6.: Simultaneous fit of Pt L3- and Co K-edge EXAFS spectra of Co_3Pt/C-800 by FEFF [75] simulated PtCo-fcc, with main fit contributions (scattering paths) for each coordination shell (C, 1, 2, 3, 4 and 5) included.

taken over as a constant for all other fits. The simulated phases do all belong to the fcc space group, thus for each path distance R_i determined in the fit, the respective lattice constant a can be derived in accord with:

$$R_n = a\sqrt{n/2} \qquad (5.1)$$

In Table 5.1 the homo- and hetero-metallic shell distances R_i ($i = 1 - 5$) are given as well as the respective lattice constants, calculated according to Eq. (5.1), for the first shell a_1 and averaged for all 5 shells $\bar{a}$. The table is divided into three block for shells visible at the Pt L3-edge only, at both edges simultaneously, and at the Co K-edge only. The results are continued in Table 5.2, where the Debye-Waller factors σ_i^2 for each shell i, simultaneously determined at both edges for Co_3Pt/C, are given, as well as the coordination numbers N_i for the homo- and hetero-metallic shells separately.

According to nano-EDX spectra that were taken from several nanoparticles visible in HAADF-STEM by H. Schulenburg et al. [81], the 'as prepared' Co_3Pt/C catalyst consists of well separated Co/C and Pt/C nanoparticles. This was confirmed by XRD measurements that they conducted and that did not show a PtCo-alloy phase for the 350 °C tempered nanoparticles. From the EXAFS fit,

5.2. EXAFS measurements of Co$_3$Pt/C nanoparticles

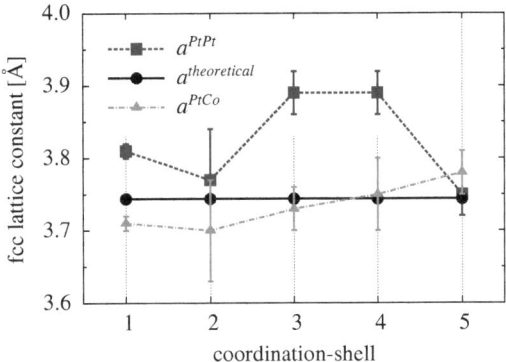

Figure 5.7.: Lattice constant a_i for Co$_3$Pt/C at 800 °C for coordination-shells i, as derived from Pt–Pt, Pt–Co, and theoretical path distances of the simultaneous EXAFS fit.

however, it can be concluded that already for Co$_3$Pt/C-asprep a small amount of Pt and Co is alloyed: $\bar{N}_i^{PtCo}$ = 0.1(1) to 0.5(4) in Table 5.2. This is feasible according to the synthesis, and since XAS measures 'all' particles and not a limited selection thereof as is done in nano-EDX. Upon tempering the samples to 350 °C and 800 °C, the Pt–Pt as well as Co–Co coordination (N_i^{PtPt} and N_i^{CoCo} in Table 5.2) is consecutively reduced, while N_i^{PtCo} is enhanced. The combined effect is most clearly visible in the decrease of the ratio quantities N^{Pt}/N^{Co} and N^{Co}/N^{Pt} at the Pt L3- and Co K-edge, respectively. It originates in the increased alloying of Pt and Co with increasing temperature, which goes along with the reduction of free Pt and Co nanoparticles. In accordance to the increase of Co atoms that penetrate into the Pt-fcc lattice, replacing Pt atoms, the Pt–Pt distances are decreasing and the Pt–Co distances are increasing towards the value 2.647 Å of the "perfect" PtCo-fcc alloy (see Table 5.1).

At 800 °C pure Co-fcc (slightly disordered) is still present in the Co$_3$Pt/C nanoparticles, since the average lattice constant is $\bar{a}^{Co}$ = 3.50(1) Å, which is even shorter as for a mono-metallic bulk Co (a = 3.54 Å). This pure Co-fcc dominates the Co K-edge EXAFS spectra and partially obscures the PtCo-fcc phase, especially the corresponding Co–Co paths. No Co-fcc was detected by nano-EDX in [81] at 800 °C, but by XRD with a = 3.544 Å. The fact that Co-fcc was not found separated from Pt via nano-EDX, can not be its rareness, but must be its attachment to PtCo alloy nanoparticles, either in the form of core and shell (with Co-fcc in the core) or via multi-crystalline domains inside one PtCo nanoparticle. At the Pt L3-edge, by contrast, no pure Pt-fcc is visible, rather the PtCo-fcc phase is clearly recognizable. Here, the Co coordination-shells are a few 0.01 Å closer compared to that of the theoretical PtCo-fcc phase, with lattice constant a = 3.7440 Å as determined by XRD [81], and the Pt coordination-shells are farther away by several 0.01 Å. This is visualized in Fig. 5.7, whereby, according to Eq. (5.1), the lattice constants are plotted instead of the path distances. Obviously, this trend is continued to about the 4th shell, but not any more to the 5th. However, beyond the 4th coordination shell, the fit quality is getting worse, as the respective spectral feature are getting weaker, in particular at the Co K-edge (see Fig 5.6). With the help of Fig. 5.7 it can be further on concluded that Co$_3$Pt/C-800 exhibits a rather disordered PtCo-fcc lattice structure, since neither the difference of a_i^{PtPt} and a_i^{PtCo} is constant, nor that of a_i^{PtPt} or a_i^{PtCo} to $a_i^{theoretical}$. As EXAFS probes all particles, regardless whether they are in a crystalline or an amorphous state, it is concluded that the dominant part of the Co$_3$Pt/C catalyst is amorphously alloyed. As a consequence, it is not possible to confirm

5. Co-Pt Nanoparticles as Catalysts in Fuel Cells

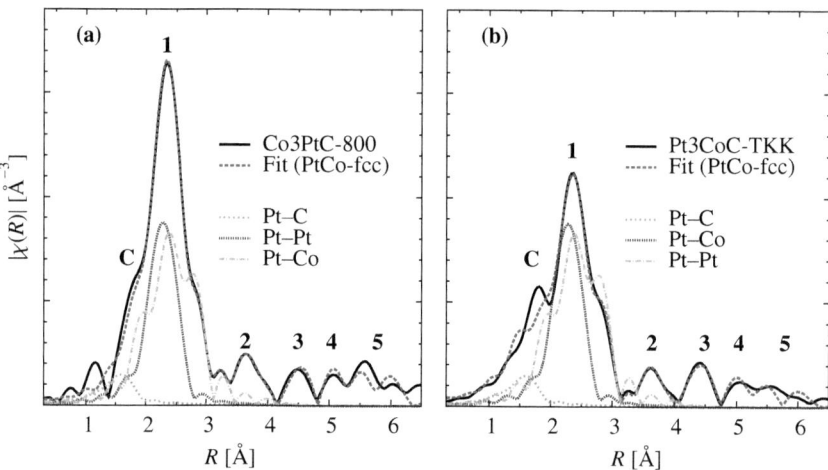

Figure 5.8.: Fit of Pt L3-edge EXAFS spectra of Co_3Pt/C-800 and Pt_3Co/C-TKK by a disordered PtCo-fcc phase. The labels C and 1 to 5 denote the coordination-shells, whereof the contributions to the first shell are shown.

or object a correlation of the various crystallographic PtCo phases found by XRD [81] with the ORR increase, as these phases are (obviously) overlaid by the amorphous part. Or, to put it another way, these PtCo phases can not be responsible for the ORR activity solely, as their contribution to the Co_3Pt/C nanoparticles is small compared to the amorphous part.

To find correlation(s) between the ORR activity, as determined by RDE measurements, and physical properties, the EXAFS spectra of Co_3Pt/C-asprep, 350 and 800 as well as of the commercial catalyst Pt_3Co/C-TKK have been fitted again, but at the Pt L3-edge only, shown in Fig. 5.8. The reason for the exclusion of the Co K-edge spectra is the aforementioned dominance of the unalloyed Co-fcc, which superposes all other phases. The fit results do resemble those from the previous EXAFS fit within the error margins, though they are more precise on average. They are listed in Table 5.3 and are (partially) visualized in Fig. 5.9. Except for the coordination number ratio N^{Pt}/N^{Co}, which was simultaneously determined for all 5 coordination shells, all values are only given for the first shell. The reason for this is that the first shell is more intense by more than a factor of 5 compared to the other shells (see Fig. 5.6) and thus yields the most reliable results, i.e. with smallest errors (see Table 5.1 and 5.2). Instead of the $Pt_{2.5}Co/C$ catalyst from ETEK utilized in [81], only a Pt_3Co/C catalyst from TKK was available for the EXAFS measurements. It resembles the former, with respect to the phase (Pt_3Co-$L1_2$, according to XRD measurements not shown), just the Co fraction is slightly lower and the metal loading higher.

In Fig. 5.9 all values, but the Pt–Co distance R_1^{PtCo}, seem to be related to the ORR activity. As there is some correlation between those values, they will be divided into two groups that can anon be identified with the two main origins of the 5d band center lowering (and thereby with the ORR activity increase), found by XANES (see Fig. 5.4):

1. Strain effect: In Fig. 5.9 (d) one can see that the interatomic Pt distance $R_1^{PtPt} = 2.688(8)$ Å is smallest for Co_3Pt/C-800. This is caused by the increased Co fraction of the PtCo-fcc alloy

5.2. EXAFS measurements of Co₃Pt/C nanoparticles

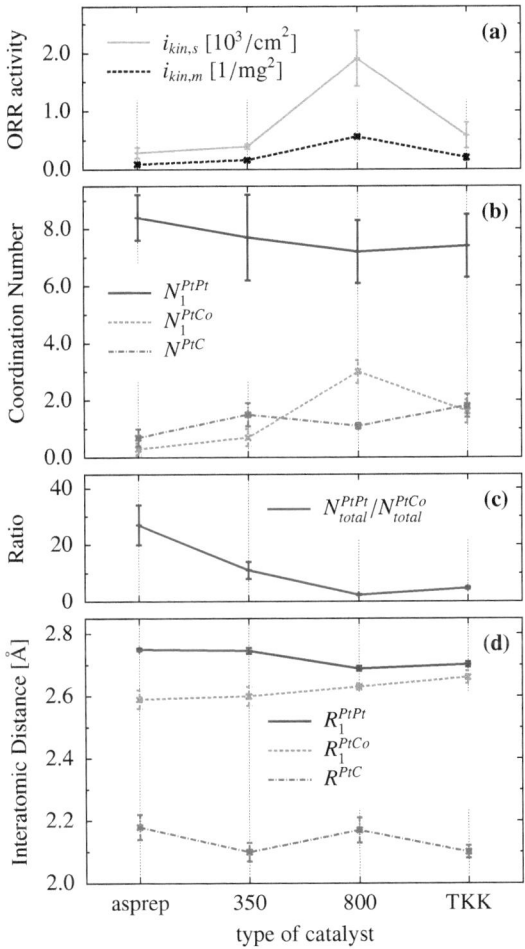

Figure 5.9.: Diffusion-corrected specific ($i_{kin,s}$) as well as mass activity ($i_{kin,m}$) for the ORR as determined by RDE measurements at 0.9 V [81] (a) in comparison to selected EXAFS fit quantities for Co₃Pt/C-asprep, 350, 800 and Pt₃Co/C-TKK (b - d).

(known as Vergard's law), visible in the parameter $N_1^{PtCo} = 3.0(4)$ and $N_1^{PtPt} = 7.2(11)$ or their ratio N_1^{Pt}/N_1^{Co} (which is identical to $N_{total}^{Pt}/N_{total}^{Co}$), shown in Fig. 5.9 (b+c).

2. Ligand effect: As can be seen in Fig. 5.9 (b+d) the Pt–C coordination N^{PtC} has a local minimum and the bond distance R^{PtC} a local maximum value for Co₃Pt/C-800. Both values are similar to those of Co₃Pt/C-asprep that, however, exhibits only a marginal amount of alloyed PtCo and therefore a rather large interatomic Pt distance of $R_1^{PtPt} = 2.749(4)$ Å, close to that of pure Pt-fcc.

49

The reason for the local extrema of N^{PtC} as well as R^{PtC} could be the tempering (compare Table 5.3): Initially, the Co nanoparticles that are part of the Co$_3$Pt/C catalyst are slightly bound to C: $N^{PtC} = 0.7(3)$. Due to the heating to 350 °C, the bondage gets tighter: doubling of N^{PtC} and shortening of R^{PtC} by about 0.1 Å. The further tempering up to 800 °C, however, loosens the Pt–C bondages (increase of R^{PtC} by about 0.1 Å) and eventually breaks them partially (decrease of N^{PtC} about 50 % again).

According to [81] Co$_3$Pt/C-600 exhibits the same Co fraction (and similar Pt loading) as Co$_3$Pt/C-800, but with 1.8 times lower ORR activity. The Pt–Co ratios have been determined by inductively coupled plasma optical emission spectrometry (ICP-OES), where element characteristic emission is detected, however, without distinguishing between the sources of that emission. Consequently, unalloyed Co and Pt contributes to the total signal as well as the alloyed parts. Thus, most probably the strain effect will not be that strong in the Co$_3$Pt/C-600 nanoparticles as less Co is alloyed to Pt, as a consequence of the lower temperature. On the other hand, the Co$_3$Pt/C nanoparticles at 900 and at 1000 °C showed lower ORR activities as well, so it seems that 800 °C just is the most suitable alloying temperature for this purpose. Unfortunately, Co$_3$Pt/C tempered to 600 °C, 700 °C, 900 °C and 1000 °C have not been available for XANES/EXAFS measurements in order to prove that hypothesis.

5.3. Conclusion

To sum up, the increase of the oxygen-reduction reaction (ORR) activity is attributed to a weakened Pt–O bond strength as a result of the lowered Pt 5d-band center, which is visible in the Pt L3-edge XANES spectrum of Co$_3$Pt/C-800. The origin of the 5d band lowering is due to the lowering of the Pt–Pt bond distance, as a consequence of the high fraction of alloyed Co (strain effect) and due to an increase of the Pt–C bond distance (ligand effect) – both visible in the EXAFS analysis. Here, the latter effect is owing to the tempering to 800 °C relative to Co$_3$Pt/C-350 and Pt$_3$Co/C-TKK. Compared to a commercial PtCo$_x$/C catalyst, Co$_3$Pt/C-800 has a (if diffusion-corrected) 3.5 and 2.8 times, respectively, higher catalytic activity as determined in [81]. Besides the ORR increase, the probability of CO poisoning should be reduced too, due to the weaker metal–oxygen bond strength resulting from the lowered 5d band (compare [45]). Moreover, 800 °C can be stated the most appropriate temperature to create this specific (at most) amorphous Pt–Co alloy.

It has to be noted that actually mainly the catalysts' surface is responsible for the ORR activity, but the results found by XANES and EXAFS are averages of the complete particles. The surface-specific values could slightly deviate from these average values, which is, however, hardly assessable. A solution would be using the technique of HRFD-XAS (introduced in section 2.3.2) that allows, in principle, for the extraction of site-selective XANES/EXAFS spectra, as will elaborated in the next chapter, though on a simpler class of nanoparticles.

5.3. Conclusion

Table 5.1.: EXAFS fits of Co_3Pt/C nanocatalysts and references at both edges. Part 1.

	perfect crystal	Foils	Co_3Pt/C asprep	Co_3Pt/C 350	Co_3Pt/C 800
Pt-L3	Pt-fcc	Pt-foil			
R-factor [$\times 10^{-3}$]	–	19	14	42	30
E_0 [eV]	–	8.1(4)	8.0(8)	8.6(11)	8.1(8)
R^{PtC} [Å]	–	–	2.17(4)	2.11(3)	2.17(4)
R_1^{PtPt} [Å]	2.772	2.758(2)	2.751(5)	2.751(7)	2.697(7)
R_2^{PtPt} [Å]	3.920	3.900(4)	3.88(3)	3.86(2)	3.77(7)
R_3^{PtPt} [Å]	4.801	4.776(4)	4.80(1)	4.79(2)	4.76(3)
R_4^{PtPt} [Å]	5.544	5.515(5)	5.54(2)	5.53(2)	5.50(4)
R_5^{PtPt} [Å]	6.198	6.166(6)	5.81(6)	5.81(10)	5.94(5)
a_1^{Pt} [Å]	3.920	3.900(4)	3.89(1)	3.89(1)	3.81(1)
$\bar{a}^{Pt}$ [Å]	3.920	3.900(2)	3.85(2)	3.85(2)	3.82(2)
Pt-L3/Co-K	PtCo-fcc	–			
R-factor [$\times 10^{-3}$]	–	–	66	73	21
R_1^{PtCo} [Å]	2.647	–	2.61(2)	2.61(3)	2.63(1)
R_2^{PtCo} [Å]	3.744	–	3.70(3)	3.65(6)	3.70(7)
R_3^{PtCo} [Å]	4.585	–	4.31(9)	4.32(6)	4.57(4)
R_4^{PtCo} [Å]	5.295	–	5.28(52)	5.07(26)	5.30(7)
R_5^{PtCo} [Å]	5.920	–	5.90(58)	5.78(37)	5.97(5)
a_1^{PtCo} [Å]	3.744	–	3.70(3)	3.69(4)	3.71(1)
$\bar{a}^{PtCo}$ [Å]	3.744	–	3.67(11)	3.62(6)	3.73(2)
Co-K	Co-fcc	Co-foil			
R-factor [$\times 10^{-3}$]	–	6	119	104	10
E_0 [eV]	–	9.7(4)	10.1(21)	11.3(15)	6.0(7)
R^{CoC} [Å]	–	–	2.00(12)	1.98(2)	1.93(2)
R_1^{CoCo} [Å]	2.506	2.501(1)	2.49(1)	2.48(1)	2.481(5)
R_2^{CoCo} [Å]	3.544	3.537(1)	3.51(6)	3.46(3)	3.43(3)
R_3^{CoCo} [Å]	4.341	4.332(1)	4.33(3)	4.36(3)	4.31(2)
R_4^{CoCo} [Å]	5.012	5.002(1)	5.03(8)	5.04(8)	4.96(3)
R_5^{CoCo} [Å]	5.604	5.592(2)	5.76(14)	5.76(15)	5.60(2)
a_1^{Co} [Å]	3.544	3.537(1)	3.52(2)	3.51(2)	3.51(1)
$\bar{a}^{Co}$ [Å]	3.544	3.537(0)	3.55(3)	3.56(2)	3.50(1)

5. Co-Pt Nanoparticles as Catalysts in Fuel Cells

Table 5.2.: EXAFS fits of Co_3Pt/C nanocatalysts and references at both edges. Part 2.

	perfect crystal	Pt/Co foil	Co_3Pt/C asprep	Co_3Pt/C 350	Co_3Pt/C 800
σ_1^2 [10^{-3}Å^2]	–	4.4(2)/6.1(2)	3.9(7)	6(1)	5.6(6)
σ_2^2 [10^{-3}Å^2]	–	7(1)/10(1)	9(3)	5(2)	11(3)
σ_3^2 [10^{-3}Å^2]	–	7(1)/11(1)	6(1)	9(2)	12(2)
σ_4^2 [10^{-3}Å^2]	–	9(2)/25(3)	17(6)	17(7)	15(3)
σ_5^2 [10^{-3}Å^2]	–	11(3)/32(12)	15(7)	18(11)	10(3)
	Pt-fcc	Pt-foil			
N^{PtC}	–	–	0.7(3)	1.6(4)	1.1(4)
N_1^{PtPt}	12.0	12.0	7.3(8)	7.5(11)	6.5(8)
N_2^{PtPt}	6.0	6.0	3.7(4)	3.8(6)	3.3(4)
N_3^{PtPt}	24.0	24.0	15(2)	15(2)	13(2)
N_4^{PtPt}	12.0	12.0	7.3(8)	7.5(11	6.5(8)
N_5^{PtPt}	24.0	24.0	15(2)	15(2)	13(2)
	PtCo-fcc	–			
$\bar{N}_1^{PtCo}$	6.0	–	0.2(2)	0.9(4)	2.5(4)
$\bar{N}_2^{PtCo}$	3.0	–	0.1(1)	0.4(3)	1.3(2)
$\bar{N}_3^{PtCo}$	12.0	–	0.5(4)	1.8(9)	5.0(7)
$\bar{N}_4^{PtCo}$	6.0	–	0.2(2)	0.9(4)	2.5(4)
$\bar{N}_5^{PtCo}$	12.0	–	0.5(4)	1.8(9)	5.0(7)
	Co-fcc	Co-foil			
N^{CoC}	–	–	1.9(3)	1.8(4)	0.9(2)
N_1^{CoCo}	12.0	12.0	4.2(6)	3.7(5)	3.5(3)
N_2^{CoCo}	6.0	6.0	1.4(2)	1.9(3)	1.7(2)
N_3^{CoCo}	24.0	24.0	5.5(7)	7.5(10)	7.0(6)
N_4^{CoCo}	12.0	12.0	2.8(3)	3.7(5)	3.5(3)
N_5^{CoCo}	24.0	24.0	5.5(7)	7.5(10)	7.0(6)
$N_{total}^{Pt}/N_{total}^{Co}$ (Pt-L3)	1.0	–	17(4)	11(3)	2.2(2)
$N_{total}^{Co}/N_{total}^{Pt}$ (Co-K)	1.0	–	∞	3.5(9)	1.8(2)

Table 5.3.: EXAFS fits of Co$_3$Pt/C nanocatalysts as well as of Pt$_3$Co/C-TKK commercial catalyst at the Pt L3-edge, in comparison to ORR activities at 0.90 V as determined by RDE measurements [81].

Pt-L3	Co$_3$Pt/C asprep	Co$_3$Pt/C 350	Co$_3$Pt/C 800	Pt$_3$Co/C TKK
R-factor [×10^{-3}]	10	40	16	21
E_0 [eV]	7.8(7)	7.7(14)	6.3(9)	6.2(10)
σ_1^2 [10^{-3}Å^2]	4.7(6)	5.8(12)	6.3(8)	7.0(9)
N_1^{PtPt}	8.4(8)	7.7(15)	7.2(11)	7.4(11)
N_1^{PtCo}	0.3(2)	0.7(3)	3.0(4)	1.6(4)
N^{PtC}	0.7(3)	1.5(4)	1.1(1)	1.8(4)
$N_{total}^{Pt}/N_{total}^{Co}$	27(7)	11(3)	2.4(2)	4.7(6)
R_1^{PtPt} [Å]	2.749(4)	2.745(9)	2.688(8)	2.701(8)
R_1^{PtCo} [Å]	2.59(3)	2.60(3)	2.63(1)	2.66(2)
R^{PtC} [Å]	2.18(4)	2.10(3)	2.17(4)	2.10(2)
a_1^{PtPt} [Å]	3.89(1)	3.88(1)	3.80(1)	3.82(1)
a_1^{PtCo} [Å]	3.67(4)	3.67(5)	3.72(2)	3.76(3)
ORR activities A				Pt$_{2.5}$Co/C
$i_{kin,s}$ [$A \times 10^3$/cm^2]	0.29(9)	0.39(4)	1.9(5)	0.6(2)
$i_{kin,m}$ [A/mg^2]	0.09(3)	0.16(2)	0.56(3)	0.20(5)

6. Site-Selective XAS

A lot of chemical synthesis routes exist to control the production of metallic nanoparticles of any desired size and shape. Various stabilizing agents and surfactants are used for that reason, and it is beyond dispute that these agents do strongly determine not only the size and shape, but also the geometrical, magnetic and electronic properties of the final product. Finally, it turned out that there is a strong interdependence of those properties, and that it is quite difficult to change one property in a systematic manner without influencing the others[37, 114].

To shed light on this complex issue of the nanoparticles interaction with its surfactants, one needs to distinguish the different sites that the metal atoms occupy inside a nanoparticle. With respect to at least the first coordination-shell, there are atoms inside the particles having the normal bulk number of neighbors, and the surface atoms that have a reduced number of neighbors – both exhibiting a valency of 0. Furthermore, there are the surface atoms that interact directly with the coating, may it be via a chemical bond or just via physi- or chemisorption and that have a valency > 0 – and what makes nanoparticles to mixed-valency compounds. By characterizing those sites of different valency separately, it should be possible to gain knowledge about the interacting forces between the coating and the "interior" that lead to the new and desired properties of nanoparticles.

However, although X-ray absorption is sensitive to the immediate environment of the absorbing atom, a standard XAS experiment just gives a spectrum that averages over all the different sites. To resolve these overlapping spectra, valency-selective XAS will be used, which is possible by utilizing highly-resolved fluorescence-detected XAS (HRFD-XAS see section 2.3.2), recorded at fluorescence peaks that are shifted in energy due to the different valencies of the absorbing element in different trapping sites. This kind of experiment was performed for the first time by M. M. Grush et al. [30] on Manganese mixed-valency complexes and (partial) valency- and site-selectivity, respectively, was achieved, though with signal-to-noise ratio being the main obstacle. P. Glatzel et al. [26] proceeded several years later on this issue, when devices as well as synchrotron sources had undergone a vast technical improvement. He successfully tested site-selective EXAFS on Prussian Blue and got pure, by means of a numerical procedure, site-selective EXAFS spectra. The processing of the (partially site-selective) HRFD-XAS spectra in a numerical procedure is mandatory to get pure site-selective spectra, as the respective emission lines are always partially overlapping (vide infra Fig. 6.1).

To establish a method for the extraction of pure valency- and/or site-selective XAS spectra, which then allows for the characterization of (arbitrary) mixed-valency materials with respect to each site, a series of measurements at the HASYLAB W1, and additional measurements at the ESRF ID26 beamline were conducted. Both beamlines provide a standard XAS setup in connection with high-resolution emission spectrometry, to allow for the performance of RIXS measurements and/or HRFD-XAS (see section 3.3 and 3.4).

At HASYLAB Co 1s3p RIXS measurements were conducted that started with a physical mixture of Co and Co(II)O powders, as a test system with known ratio, and continued with the investigation of smoothly oxidized cobalt nanoparticles. To refine the results obtained at HASYLAB, those investigations were continued on equally synthesized Co nanoparticles at the ESRF.

6. Site-Selective XAS

6.1. General strategy for site-selective XAS

In order to obtain real valency and/or site-selective XAS spectra of a (unknown) mixed-valency compound X, appropriate homovalent model compounds (MC) are necessary, so that the following tasks can be accomplished (description for the case of two MCs and hence two main valencies/sites):

1. Linear combination fit (LCF) of the respective valence-sensitive emission spectrum ($K\beta_{1,3}$, $K\beta_{2,5}$, etc.) of X by its MCs, which yields the MCs fractions c_{exp}^1 and c_{exp}^2 in X for each energy step. The result will serve for finding emission energy positions with highest, lowest and intermediate contrast of the MCs. Here it is desirable to have an overall uniform distribution of the MCs ratios at these positions, which could be fulfilled also upon neglecting the intermediate contrast position, i.e. the latter is not mandatory but increases the statistics of the results.

2. HRFD-XAS spectra for X and the MCs have to be obtained at these emission energy positions, either by extraction from a RIXS plane or by recording them upon detection of fluorescence from these positions solely. Note that the number of HRFD-XAS spectra (equal to number of positions in 1.) gives the maximum of accessible valencies/sites.

3. A check of the suitability of the chosen positions with respect to lifetime broadenings is provided, upon comparing the average of the MCs HRFD-XAS spectra from the chosen positions with the ones recorded at the peak (see section 2.3.3).

4. Appropriate MCs provided, a LCF of the HRFD-XAS spectra of X may be performed optionally, to have a comparison to the emission LCF, since the sensitivity of emission and absorption spectra with respect to the MCs is generally different.

5. Each HRFD-XAS spectrum S_{exp}^i ($i = 1, 2, 3$) can be written as a linear combination of the required "theoretical" site-specific spectra S_{th}^k (for the case of $k = 2$):

$$S_{exp}^i = c_{th}^{i,1} S_{th}^1 + c_{th}^{i,2} S_{th}^2, \qquad (6.1)$$

with to be determined ratio coefficients $c_{th}^{i,k}$. These three equations have to be solved by a least squares fit which will be realized by singular value decomposition (SVD) as described in appendix B and which yields the following:

- An (infinite) set of mathematical pairs of (XAS) spectra S_{th}^1 and S_{th}^2 which are only restraint by (a) having to be positive for each energy step and (b) having corresponding ratio coefficient pairs $c_{th}^{i,1}$ and $c_{th}^{i,2}$ that must sum up to one.
- By variation of the SVD parameters the theoretical spectra S_{th}^k and simultaneously the coefficients $c_{th}^{i,k}$ are varied. In this way, the coefficients are adjusted to the ratios $c_{exp}^{i,k}$ determined at point 1. or 4. to get, at least, a limited set of pairs of (physical) spectra.

6. The final set of site-specific spectra is compared to the spectra of appropriate references and simulated spectra, if available, so that further exclusion of (unphysical) spectra is possible and at best one specific spectrum for each valency and/or site is found.

6.2. 1s3p-RIXS at wiggler beamline W1

The RIXS experiments were performed at the wiggler beamline W1 of the 4.45 GeV storage ring DORIS III at HASYLAB (see section 3.3 and Table 3.2 for details). The energy of the incident X-rays

6.2. 1s3p-RIXS at wiggler beamline W1

was tuned through the cobalt absorption K-edge (7709 eV) via a double-crystal Si(111) monochromator (with resolution $\Delta E \approx 2$ eV), from 7700 eV to 7750 eV in 1 eV steps and from 7750 eV to 7780 eV in 2 eV steps. At each step 3 or 4 emission spectra have been recorded with collection time of 40 sec. For the first experiment, the emitted X-rays were detected in dispersive geometry by the high-resolution Johann spectrometer that was equipped with a spherically bent Si(531) analyzer crystal. The Bragg angle for the main $K\beta_{1,3}$ peak at 7649.4 eV was $\theta = 61.1°$. The width of the CCD chip and the energy dispersion of the crystal allowed detecting an emission energy range of 155 eV around the $K\beta_{1,3}$ line of cobalt, with energy resolution of about 1 eV. For the second experiment a Si(620) analyzer crystal was used, which leads to a Bragg angle for the main $K\beta_{1,3}$ peak of $\theta = 70.7$ °C and allowed for the detection of an energy range of 75 eV around the $K\beta_{1,3}$ line.

6.2.1. Co-CoO test system

1s3p-RIXS maps of cobalt (Co), cobalt(II)-Oxide (CoO) and a mixture thereof (Mix) as well as a cobalt(III)-Oxide (Co_2O_3), all in powder form, have been measured. The ratio for the mixture of about 80 : 20 (precisely 78.1 % Co and 21.9 % CoO) was chosen as to model a Co nanoparticle with a thin protective oxidized shell.

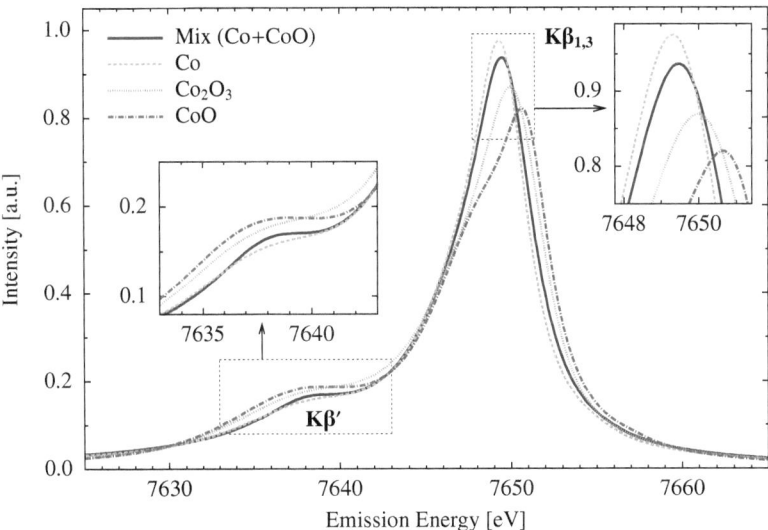

Figure 6.1.: Normalized $K\beta_{1,3}$ emission spectra of Co, CoO, Co_2O_3 and the Mix (78.1 % Co and 21.9 % CoO), all measured as powders. The insets show a magnification to both peaks $K\beta'$ and $K\beta_{1,3}$.

Off-resonant emission spectra (NRXES)

At first off-resonant $K\beta_{1,3}$ emission spectra of all samples were recorded at a fixed excitation energy of 8000 eV. To get smooth data curves as well as precise positions of the peaks, Voigt fits were performed

6. Site-Selective XAS

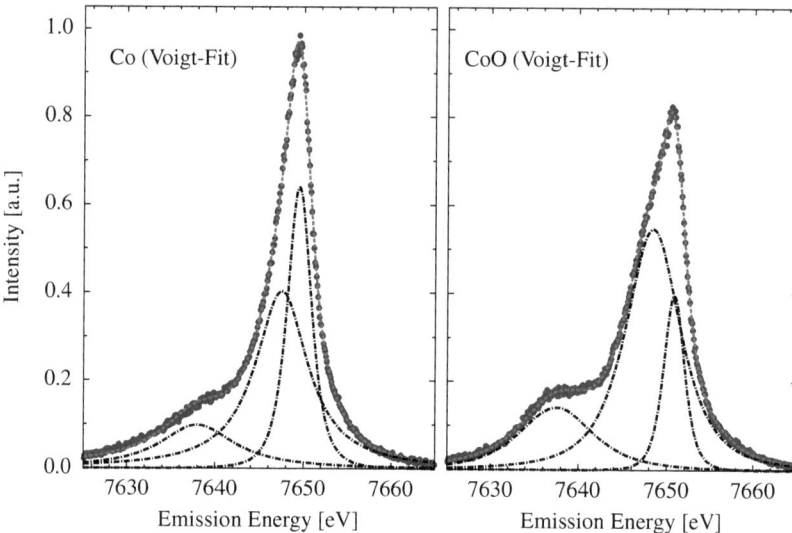

Figure 6.2.: Normalized and aligned $K\beta_{1,3}$ emission spectra (blue filled circles) of powdered Co (left) and CoO (right) fitted by three Voigt functions (black dash-dotted curves and total fit as red dashed curve) which do not represent real resonances. Fit details are given in Table 6.1.

on all four spectra with full width at half maximum (FWHM) variable between 0 and 10 eV. It has to be pointed out here that the three Voigt functions do not correspond to the real resonances (there are much more), but give a good approximation of the main contributions. The resulting $K\beta_{1,3}$ fitting curves of the Mix, Co, CoO and Co_2O_3 are shown in Fig. 6.1, along with zooms of the two peak regions $K\beta'$ at about 7638 eV and $K\beta_{1,3}$ at about 7649 eV. The spectra have been normalized with respect to the integrated (spectral) area under the curve. This area is proportional to the total 3p→1s transition probability and has to be equal for all Co compounds. Moreover, the spectra are aligned according to their first moment (the areas centroid) to compensate for any decalibrations during the measurements.

The Voigt fits of the two most distinct samples Co and CoO are shown in Fig. 6.2. The complete fitting results are listed in Table 6.1. From the positions and resulting splitting between $K\beta'$ and $K\beta_{1,3}$, the previously made statement can be verified: With increasing net valence spin (CoO > Co_2O_3 > Mix > Co), the positions of $K\beta'$ and $K\beta_{1,3}$ are shifted to lower and higher energies, respectively, so that the respective splitting, which is a measure of the 3p–3d exchange interaction, is increased.

To find positions with most distinct Co to CoO contrast and to check the physical ratio of the Mix, a linear combination fit (LCF) of the $K\beta_{1,3}$ spectrum of the Mix by those of its components Co and CoO is performed. It resulted in a ratio of 76.1 : 23.9 with error ±0.3 for each value, and a fit quality of $Q = 0.02$ and is given in the upper panel of Fig. 6.3. The fit quality is the sum of squares of residuals, i.e.

$$Q = \sum_{E}(f_E - y_E)^2,$$

6.2. 1s3p-RIXS at wiggler beamline W1

Table 6.1.: Results for the 3-Peak-Voigt-Fit of the four $K\beta_{1,3}$ emission spectra. Each column lists the energetic positions of the three Voigt-functions. The corresponding position at the $K\beta_{1,3}$ line is denoted in brackets. $K\beta_{1,3}^{eff}$ is the peak of the whole Voigt-fit, i.e. of the sum of the three Voigt-functions and ΔPeaks is the energetic difference between Peak-eff and Peak-1, i.e. the $K\beta_{1,3}$ to $K\beta'$ splitting.

	Peak-1 [eV] ($K\beta'$)	Peak-2 [eV] ($K\beta_{1,3}$-shoulder)	Peak-3 [eV] ($K\beta_{1,3}$)	Peak-eff [eV] ($K\beta_{1,3}^{eff}$)	ΔPeaks [eV] (eff↔1)
Co	7638.0	7648.1	7649.6	7649.3	11.3
Mix	7638.0	7647.9	7649.8	7649.5	11.5
Co_2O_3	7637.7	7647.8	7650.3	7650.0	12.3
CoO	7637.5	7648.5	7650.9	7650.7	13.2

where f_E and y_E is the discrete fit data and experimental data, respectively, and E is the X-ray energy. The obtained ratio deviates about 2.5 % from the physical mixing ratio 78.1 : 21.9 and therefore is outside the fitting error of ±0.5 %. The reason for this deviation could originate in self-absorption effects as a result of too thick samples. As a consequence, not all incident X-rays would leave the samples, which actually is irrelevant for non-resonant X-ray emission as there is no correlation between incident and emitted X-rays. However, it is relevant in this case as there are two components involved (Co and CoO) that have different densities (8.9 g/cm^2 and 6.4 g/cm^2) and thus different absorption lengths l_{abs}: Calculations by the software Hepheastus [69] give l_{abs} of CoO about twice that of Co. Consequently, self-absorption is weaker for CoO leading to, in this case, a slight enhancement of this component in agreement to the results.

HRFD-XANES spectra from RIXS

To find appropriate positions for the extraction of HRFD-XANES spectra, the fractions of Co and CoO in Mix, respectively, are given in the lower panel of Fig. 6.3. In order to improve the signal-to-noise ratio of these spectra, it will not be just searched for particular energies but intervals. This led to quite big intervals in particular for the non-peak regions, however, with the restriction of maintaining an almost constant ratio of the two reference compounds throughout the whole interval. Two regions were chosen where the Co to CoO contrast is lowest (Pos-3) and highest (Pos-2) and one where it is intermediate (Pos-1). The details of these intervals are listed in Table 6.2, and its mean energies are indicated by arrows in Fig. 6.3.

The normalized HRFD-XANES spectra of the Mix extracted from these intervals (step 2) are shown in the left panel of Fig. 6.4. For details about the normalization of XANES spectra, see appendix C. The most significant variations are at the edge and the whiteline at about 7712 eV and 7726 eV (in-

Table 6.2.: Details about $K\beta_{1,3}$ fluorescence intervals chosen for the extraction of HRFD-XANES spectra. ΔE is the width of the chosen intervals.

interval	$E_{interval}$ [eV]	ΔE [eV]	$K\beta_{1,3}$-LCF
Pos-1	7645.50 - 7647.51	2.01	78.2(8) : 21.8(6)
Pos-2	7648.52 - 7649.52	1.00	82.0(4) : 18.0(1)
Pos-3	7651.35 - 7653.08	1.73	64.2(8) : 35.8(10)

6. Site-Selective XAS

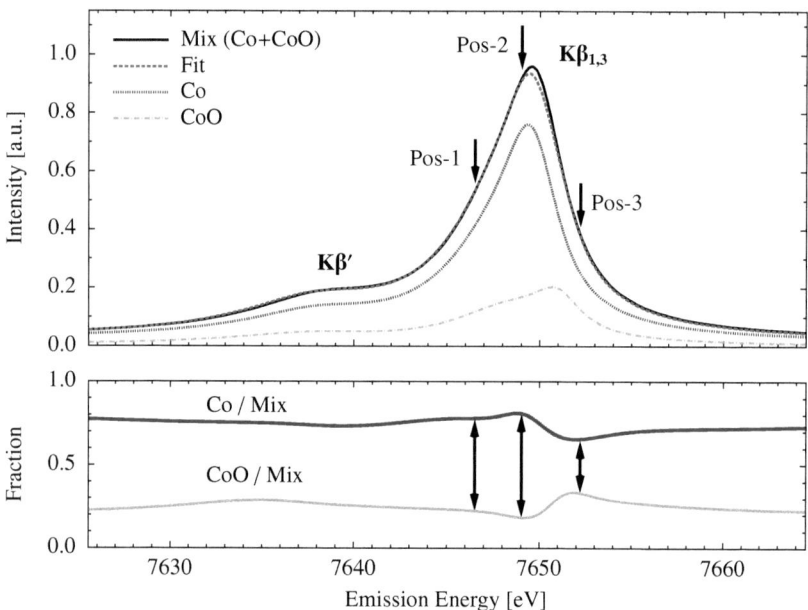

Figure 6.3.: Top: Normalized K$\beta_{1,3}$ NRXES spectra of powdered samples Mix, Co and CoO. The latter two spectra's intensities are weighted with respect to their contribution in Mix, as determined by the LCF. Bottom: Fractions of Co and CoO in Mix, as calculated by the LCF. The arrows (top and bottom figure) denote the mean energies of the intervals with intermediate (Pos-1), highest (Pos-2), and lowest (Pos-3) Co to CoO ratio, as chosen for the extraction of the HRFD-XANES spectra.

dicated by arrows), respectively, and at the shape resonances at about 7742 eV and 7760 eV. Those variations can be explained upon comparing the Mix spectrum with those of its components Co and CoO, shown in the right panel of Fig. 6.4 (only Pos-1 is shown): Decreasing intensity of the edge shoulder (7712.5 eV) and increasing intensity of the whiteline (7725.5 eV), as well as appearance of a first shape resonance at about 7742 eV and the shift to higher energies of the Co shape resonance to about 7773 eV, can all be attributed to a CoO increase (compare Fig. 5.2 in section 5.1 and explanations given there).

A further LCF was performed, this time of the HRFD-XANES spectra of the Mix by its components Co and CoO, which is shown in Fig. 6.5. Here the Mix spectrum is fitted by its components Co and CoO extracted from the same interval in each case. The quality of XANES fits is expressed by the R-factor, which is the total relative deviation between fit data and experimental data:

$$R = \frac{\sum_E (f_E - y_E)^2}{\sum_E y_E^2},$$

where f_E and y_E is the discrete fit data and experimental data, respectively, and E is the X-ray energy, i.e. the excitation energy. The resulting Co : CoO ratios are given in Table 6.3, in the column labelled

6.2. 1s3p-RIXS at wiggler beamline W1

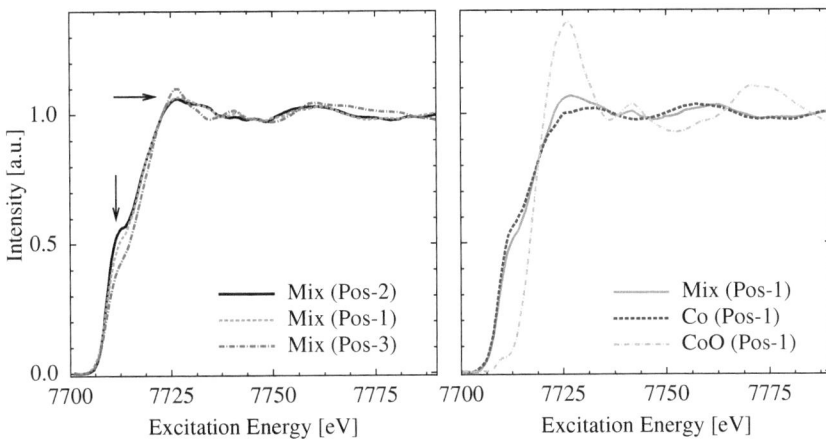

Figure 6.4.: Normalized Co K-edge HRFD-XANES spectra. Left: The Mix, extracted from fluorescence regions denoted in brackets (and indicated by arrows in Fig. 6.3). The two arrows indicate the regions that are most sensitive to the Co valency, i.e. Co^0 and Co^{2+} in this case. Right: Comparison of the Mix spectrum with those of its components Co and CoO, all extracted from the interval at Pos-1.

XANES-LCF. It seems that the Co contribution with respect to XANES is more dominant compared to $K\beta_{1,3}$ emission: An increase of about 10 % is . The reason for this lies in the different sensitivities of XAS and XES with regard to the unoccupied and occupied density of states, respectively. Anyway, the trend is the same as in the $K\beta_{1,3}$-LCF, i.e. increase of CoO fraction from Pos-2 to Pos-1 to Pos-3. Finally, it can be stated that by choosing different $K\beta_{1,3}$-LCF fluorescence energies, in due consideration of the ratios of the components known from LCF of the emission lines, the valency sensitivity is tuned and one obtains partially valency-selective HRFD-XANES spectra.

The sought pure (valence-selective) Co and CoO spectra should resemble the average spectra of Co and CoO from the three intervals at Pos-1, 2 and 3. Consequently, prior to step 3 a LCF of the HRFD-XANES spectra of the Mix by these average spectra is performed. The resulting ratios are shown in Table 6.3 in the column labelled XANES-LCF2. Within the errors they are in agreement with the results from the first XANES-LCF (listed in the same Table), albeit the fit quality is on average worse about a factor two, which is an effect of the different lifetime influences onto the three interval positions (see section 2.3.3) that are cancelled out due to the averaging for XANES-LCF2. Lastly, the averaged spectra of the components Co and CoO have to be compared to their respective HRFD-XANES spectra extracted from the respective $K\beta_{1,3}$ peak positions (precisely an interval ±1 eV around the peak), to assure the suitability of the chosen intervals with respect to the lifetime disturbances (see 2.3.3). This is given in Fig. 6.6 and shows almost perfect agreement between the average spectrum and the one from the peak, as required.

In the following (step 3) the pure valency-selective spectra of Mix shall be determined, although they are already known, as testing case for materials where the pure compounds are a priori unknown. The three ratios $c^i_{Co/CoO}$ as well as the two pure spectra S_{Co} and S_{CoO} will be treated as variables to be determined in a least squares fit (LSF). For this purpose the experimental HRFD-XANES spectra

6. Site-Selective XAS

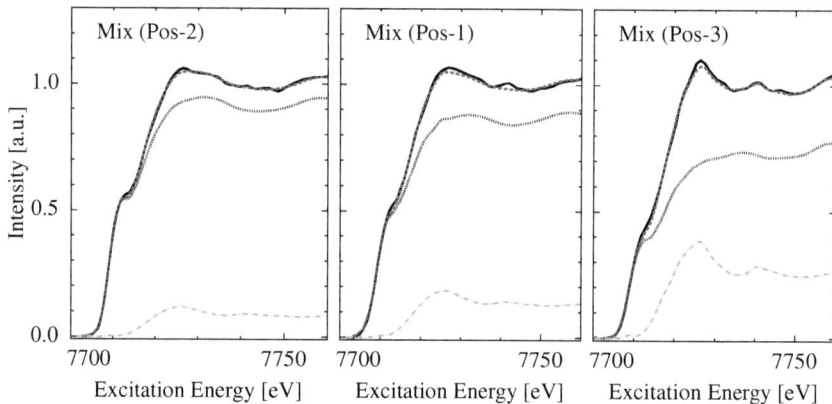

Figure 6.5.: LCF of the three HRFD-XANES spectra of the Mix. For each fit the data (black line) is shown along with the fit (red slashed) and the fit components Co (blue dotted) and CoO (light-blue dash-dotted). The two fit components are taken from the same intervals as Mix for each fit.

S^i_{exp} of the Mix will be written in dependence of the c^i ($i = 1 - 3$) and $S_{Co/CoO}$ in the following general form (compare Eq. 6.1):

$$S^i_{exp} = c^i_{Co} S_{Co} + c^i_{CoO} S_{CoO}. \tag{6.2}$$

The LSF will be performed by employment of the SVD to the matrix M_{exp} composed of the three HRFD-XANES spectra, as explained in [84] and in the appendix B.

One ends up with a set of mathematical solutions S_{Co} and S_{CoO} to Eq. (6.2), each of which being an equally good result of the least squares fit, and from whom one can in principle find the physical solutions, i.e. the pure spectra of Co and CoO. For this purpose, the coefficients $c^i_{Co/CoO}$ are adjusted to the LCF ratios (Table 6.3). It turns out that only the $K\beta_{1,3}$-LCF ratios are reproducible, as can be seen in Table 6.3. The larger Co fractions of XANES-LCF(2) would have forced the pure CoO spectrum to become unphysical and the pure Co spectrum to become less similar to the average Co spectrum.

Table 6.3.: Results of all linear combination fits (LCF) of the Mix by Co and CoO. $K\beta_{1,3}$-LCF ratios from Table 6.2 with standard deviation given in brackets. XANES-LCF: Each Mix spectrum was fitted by Co and CoO extracted from the same interval. XANES-LCF2: Each Mix spectrum was fitted by the same Co and CoO spectra which are merges from the three intervals. Last column shows ratios as fitted by SVD. For the latter three fits the fitting errors are given in brackets.

interval	$K\beta_{1,3}$-LCF	XANES-LCF	XANES-LCF2	XANES-SVD
Pos-2	82.0(4) : 18.0(1)	91.4(3) : 8.6(3)	93.7(11) : 6.3(11)	84.1(3) : 15.9(3)
Pos-1	78.2(8) : 21.8(6)	86.7(5) : 13.3(5)	86.8(7) : 13.2(7)	77.1(3) : 22.9(3)
Pos-3	64.2(8) : 35.8(10)	73.8(6) : 26.2(6)	74.7(8) : 25.4(8)	61.1(2) : 33.9(2)
R-factors [e-6]		55, 113, 133	501, 249, 210	40, 50, 13

6.2. 1s3p-RIXS at wiggler beamline W1

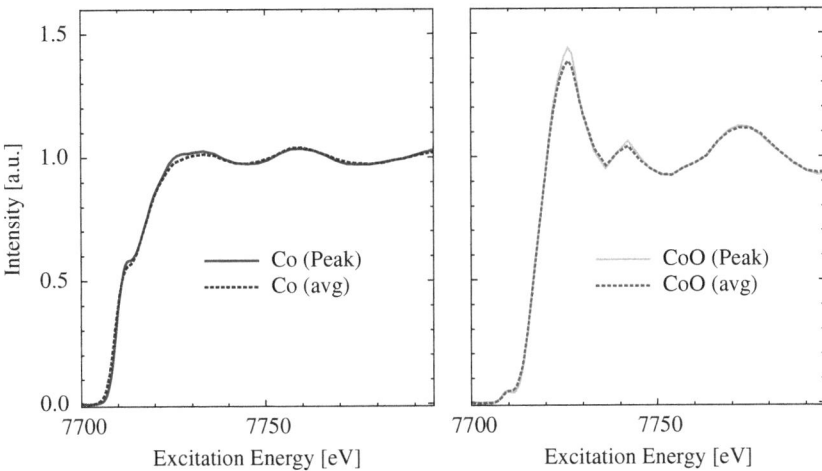

Figure 6.6.: Normalized HRFD-XANES spectra of Co and CoO extracted at the $K\beta_{1,3}$ emission peak (peak), in comparison to the average of the HRFD-XANES spectra extracted from the three intervals (avg).

The final pure spectra of Co and CoO are shown in the left panel of Fig. 6.7, in comparison to the averaged HRFD-XANES spectra (merges of the spectra from the three positions). Obviously, it was possible to find the spectrum for Co, except for slight intensity exaggerations, but not for CoO that shows some significant differences with respect to energetic positions, e.g., low-energetic pre-edge and both shape resonances with wrong centroids.

It is possible that the ratio of CoO in Mix is too small to be representative enough, to be identifiable in the general least squares fit via SVD. In order to improve the weight of the CoO contribution, the interval at Pos-3, which have the highest CoO fraction, will be expanded and split to obtain two CoO-rich HRFD-XANES spectra. Then the SVD is applied again according to Eq. (6.2) onto four experimental spectra (Pos-1, Pos-2, Pos-3a and Pos-3b), i.e. $i = 4$ this time. The results obtained by this modified SVD, however, show no improvement. One last simplification is tried: The SVD is applied with the final spectrum S_{Co} fixed to Co-avg from the beginning to calculate S_{CoO} and then with S_{CoO} fixed to CoO-avg to calculate S_{Co}. Both resulting spectra are given in Fig. 6.7 in the right panel. The Co spectrum could be reproduced almost perfectly, but the CoO spectrum shows still some significant differences.

There are several reasons for the inappropriate reproducibility of CoO. From the mathematical point of view, it is not necessary for the theoretical CoO spectrum to have, e.g., a pre-edge at exactly 7709.5 eV, however, if one HRFD-XANES spectrum of the Mix had shown the pre-edge, it would have been reproduced by the SVD. It can be inferred hence, that a sensitivity limit of the SVD has been encountered: The spectra to be found by SVD have to be significantly present inside the multivalent compound, which implies that their finest spectral structures already have to show up in the HRFD-XANES spectra of the multivalent compound, otherwise they will not be reproduced reliably. In conclusion hence, the strategy to achieve valency-/site-selective XANES spectra of a testing system have been only partly successful. Nonetheless, it will be applied to real multivalent samples – Co

6. Site-Selective XAS

nanoparticles – since they reveal themselves to be a mixture of about 55 : 45 of metallic to divalent Co, so that one is far away from the sensitivity limit found in this section. Furthermore, the data quality of the following measurements is much better, so that it can be presumed that much more fine structures will be already visible in the nanoparticles' HRFD-XANES spectra.

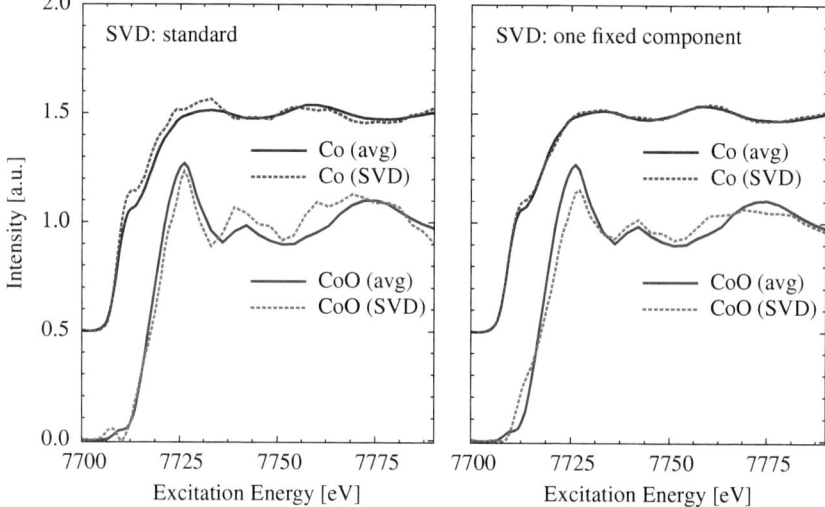

Figure 6.7.: Normalized Co K-edge spectra of Co and CoO as obtained from the SVD in comparison to their experimental HRFD-XANES spectra Co and CoO ("avg" denotes the average of the three HRFD-XANES spectra).

6.2.2. Co nanoparticles

This section was completely published in [49]. The 1s3p-RIXS maps of Co nanoparticles (Co-nano) as well as fully oxidized Co nanoparticles (Co-nano-ox) (both described in section 4.1.1) and a standard 7.5 μm thick metallic Co foil (Co-foil) have been measured. According to the Co nanoparticles' synthesis (section 4.1.1 and [7]) and previous studies [77], the main sites are assumed to be a protective shell of CoO/ $CoCO_3$ (valency +2) and a metallic Co core (valency 0) with as yet unknown crystal structure. This nanoparticle is shown schematically in Fig. 6.8. Due to this assumption of well separated core and shell sites, the valency-selectivity likewise is a site-selectivity.

To clarify the electronic and geometric structure of both sites, it will be advanced according to the elaborated strategy (section 6.1). First, a LCF of the Co-nano $K\beta_{1,3}$ line (LCF-$K\beta_{1,3}$) is performed, with the two references Co-nano-ox and Co-foil that are taken as model compounds. The fit resulted in an overall Co-nano-ox : Co-foil ratio of 44 : 56 with errors ±1 and is shown in Fig. 6.9 (top). At first, the Co-nano spectrum can be classified in between Co-foil and Co-nano-ox so that according to the interpretation of the $K\beta_{1,3}$ to $K\beta'$ splitting as a measure for the net valence spin (see section 2.3.2), it has to lie in between the values of those two references. According to the obtained ratio, one can estimate the scales of the simple core-shell model (see appendix A). Therefore, the supposition of spherically shaped nanoparticles will be made, which is strongly favored with respect to the synthesis

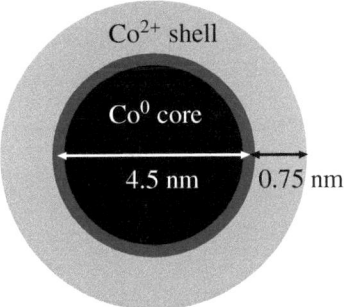

Figure 6.8.: Simple spherical core-shell model. A metallic Co core surrounded by a CoO/C shell of valency 2 and in between a transition layer.

too. However, other shapes are conceivable as well so that the results based on this supposition are to be dealt with care. By using the bulk values for the density of Co and CoO one gets a core diameter of ≃ 4.5 nm and a shell thickness ≃ 0.75 nm. The latter implies that about 2 to 3 layers of CoO are covering the core. Coming back to the fit (Fig. 6.9), it is obviously not possible to reproduce both $K\beta$ peaks correctly. This indicates that the sought sites are not identical to an oxidized Co-nano and/or a bulk metallic Co-foil and/or that more than two main sites are present. To be precise, there is a surface layer of the Co core adjoining the inner surface of the CoO shell and both surfaces exhibit a reduced number of neighbors, but have the same valencies as the "bulk" core and shell, respectively. In between those adjoining surfaces some transition layer exists, comprised of both Co valencies. The shell's outer surface could have reduced neighbors too and could be connected to remnants from the synthesis (C, N or Al) (see Fig. 6.8 for visualization of all sites). However, since the shell only possesses about 2 to 3 layers, its surfaces are the dominant (or even only) contributions. Therefore, there are effectively three different Co valencies/sites present: (1) the core with its surface, (2) the transition layer and (3) the (thin) shell. The transition layer will be shown to be negligible, in comparison to the main sites, in the following as it is overlaid by the noise level. Furthermore, although Co^0 is presumed to build the core and Co^{2+} the shell, mixing of those valencies in between these two sites is possible – e.g., in the form of several more or less crystalline areas of Co^0 surrounded by Co^{2+} inside the nanoparticle – and can not be identified by the utilized method.

The fractions of the two model compounds, as obtained by the fit, are shown in Fig. 6.9 (bottom).

6. Site-Selective XAS

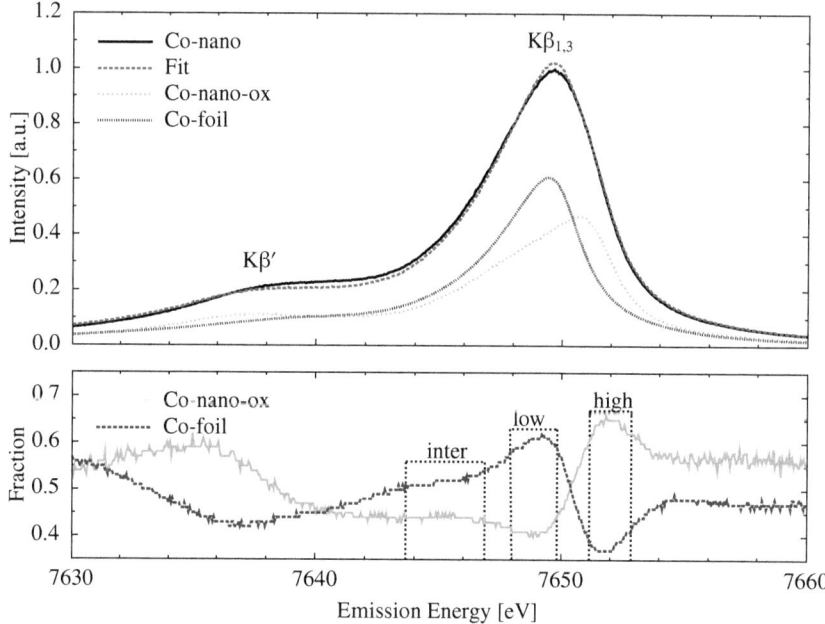

Figure 6.9.: Top: $K\beta_{1,3}$ emission lines of Co-nano, Co-nano-ox and Co-foil. The latter two spectra are scaled with respect to their contribution in Co-nano, as calculated by the LCF. Bottom: Fractions of Co-nano-ox and Co-foil in Co-nano, as calculated by the LCF. Intervals (shown as dashed boxes) denote regions with intermediate, lowest and highest shell to core ratio.

They serve for finding appropriate emission energy intervals (again to improve signal-to-noise ratio) for the extraction of the HRFD-XANES spectra of the Co nanoparticles. Those intervals are chosen big enough to maintain a strong signal on the one hand and small enough to keep the ratio of the two model compounds almost constant throughout the whole interval on the other hand. Two regions are selected, where the shell to core contrast is lowest (low) and highest (high), respectively, and one where it is intermediate (inter). These intervals are shown in Fig. 6.9 (bottom) as dashed boxes. The explicit mean ratios of these intervals are given in Table 6.4 (column labeled LCF-$K\beta_{1,3}$), with standard deviation of ±1 for each value and errors similar to those of the overall fit ratios.

The extracted HRFD-XANES spectra are shown in Fig. 6.10. The partial site-selectivity of these spectra is visible in the variation of the edge shoulder and the whiteline at about 7713 eV and 7725 eV, respectively. As in the previous section (see Fig. 6.4), both effects, decrease of edge shoulder and increase of whiteline (as well as shift of the first shape resonance at 7760 eV), can be understood as an increase of the Co^{2+} fraction (compare right panel of Fig. 6.10).

To cross-check the ratios from LCF-$K\beta_{1,3}$, a LCF of the HRFD-XANES spectra is performed too. The Co-nano and the reference compounds were taken from the same intervals to perform the fits, i.e. Co-foil (high) and Co-nano-ox (high) for Co-nano (high) etc. The results are shown in Table

6.2. 1s3p-RIXS at wiggler beamline W1

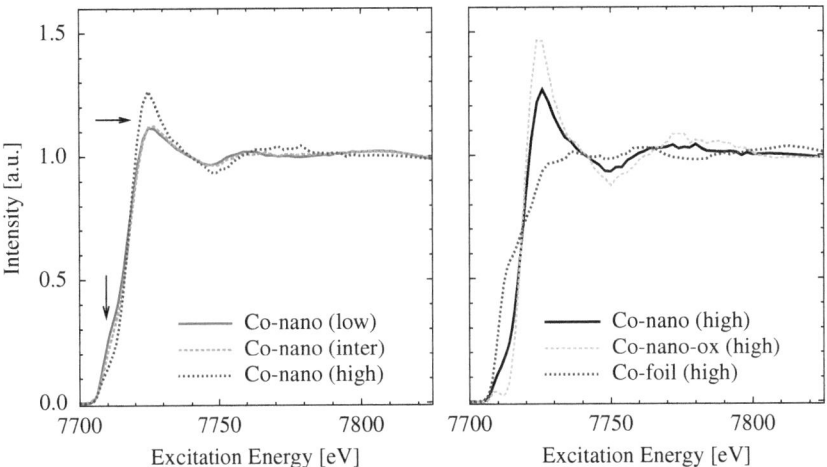

Figure 6.10.: Normalized Co K-edge HRFD-XANES spectra. Left: Co-nano, extracted from fluorescence regions (shown in 6.9 bottom) with most distinct shell rates as denoted in brackets. The two arrows indicate the regions that are sensitive to the Co valency. Right: Co-nano and its model compounds extracted at the "high" interval.

6.4 in the column labelled LCF-XANES, and they almost coincide with the ratios from the previous $K\beta_{1,3}$-LCF.

Before starting to calculate the pure site-selective spectra, the suitability of the chosen intervals is checked, as elucidated in section 2.3.3. For this purpose, the model compounds average of the three HRFD-XANES spectra extracted from the intervals, are compared to the HRFD-XANES spectra extracted from an interval (±1 eV) around the respective $K\beta_{1,3}$ emission peak (7649 eV for Co-foil and 7651 eV for Co-nano-ox) in Fig. 6.11. Each pair of spectra shown must be at best equal, which is almost perfectly the case, so that the lifetime disturbances do not further have to be taken into account.

To get real site-selectivity, the pure spectra $S_{core/shell}$ are extracted from the experimental ones S^i_{exp},

Table 6.4.: Ratios Co-nano-ox : Co-foil in Co-nano, as obtained by LCF of the $K\beta_{1,3}$ line (standard deviation and fitting errors ±1) and as obtained by LCF of HRFD-XANES spectra (fitting errors ±0.5). Last two columns: Ratio shell : core in Co-nano as calculated by SVD: first the minimum and maximum ratios (with respect to the shell fraction), then the average ratios (errors negligible small about ±0.1).

interval	LCF-$K\beta_{1,3}$	LCF-XANES	SVD (min / max)	SVD (average)
high	63 : 37	63 : 37	65 : 35 / 81 : 19	73 : 27
inter	46 : 54	48 : 52	43 : 57 / 51 : 49	47 : 53
low	40 : 60	43 : 57	36 : 64 / 45 : 55	40 : 60
R-factor [e-5]	95	27, 10, 15	–	4, 10, 8

6. Site-Selective XAS

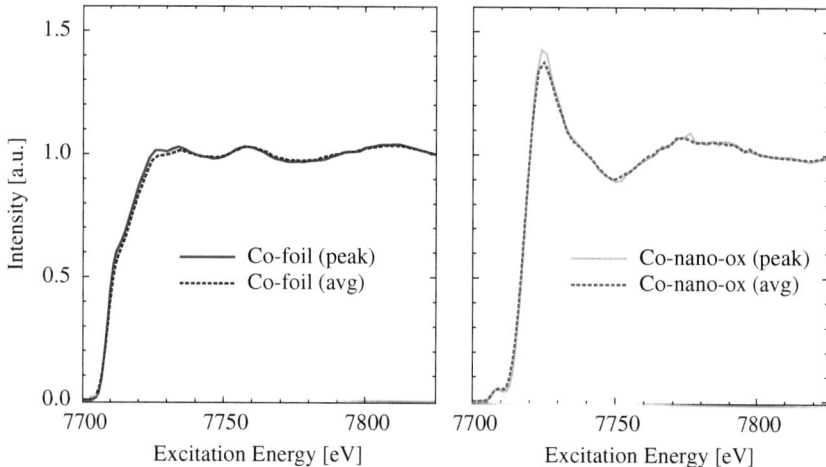

Figure 6.11.: Normalized HRFD-XANES spectra of the model compounds extracted at the $K\beta_{1,3}$ emission peak (peak), in comparison to the average of the HRFD-XANES spectra extracted from the three intervals (avg).

and thus the experimental HRFD-XANES spectra are written as (compare Eq. 6.1)

$$S^i_{exp} = c^i_{core} S_{core} + c^i_{shell} S_{shell}, \qquad(6.3)$$

with the coefficients $c^i_{core/shell}$ and $i = 1 - 3$. Since neither the exact number of main sites inherent in the nanoparticles is known, nor the exact values for the coefficients, the number of theoretical sites will be extended to the maximum of 3 (given by the number of input files): S_{core}, S_{shell} and S_{other}. This serves as a check on the required number of main sites.

Upon solving Eq. (6.3) with the help of a SVD (see appendix B), the first result gained is that only two pure spectra are necessary as basis to reproduce M_{exp} (the matrix composed out of the three HRFD-XANES spectra) and are distinguishably from noise, respectively. This implies that only two main sites are necessary, as was proposed. To find the physical solutions from the set of mathematical ones, it is demanded that the coefficients are at least similar to the fractions of the model compounds, as obtained by LCF-$K\beta_{1,3}$ and LCF-XANES, respectively (see Table 6.4), for each experimental spectrum.

One ends up with a limited range for the SVD parameters and therefore with a small set of physical solutions shown in Fig 6.12. The respective calculated ratios are listed in Table 6.4 along with those from the LCF's. Only the extreme cases are shown and each arbitrary ratio-triple in between those extremes is possible. The LCF ratios for the intermediate- and the lowest-shell spectrum of Co-nano are nearly equal to the average values of the calculated ones from the SVD. The ratio for the high-shell spectrum by contrast, could not be fitted in the same way. Not even the minimum of the SVD calculation reaches the low shell fraction of the respective LCF-$K\beta_{1,3}$ result. This again is an indication for the insufficient representation of the real shell by Co-nano-ox and/or the real core by Co-foil.

The average core spectrum, with its standard deviation, in comparison to the Co-foil spectrum, is

6.2. 1s3p-RIXS at wiggler beamline W1

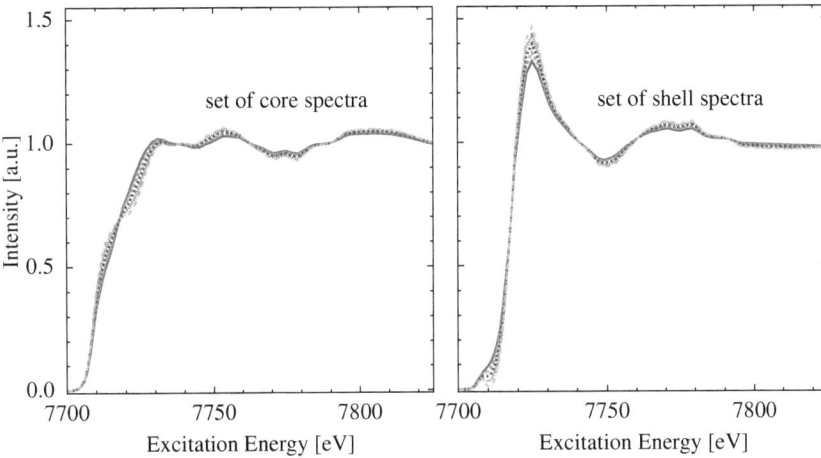

Figure 6.12.: Final set of site-selective XANES spectra for the two sites, core and shell, of the Co nanoparticles, as obtained from the numerical procedure and in consideration of the restraints. Five equidistant steps in between the final parameter range are applied for both sites.

given in the left panel of Fig. 6.13. The core shows significant differences to Co-foil – a mixture of hcp and fcc, see appendix C – especially with respect to the shape of the edge and the whiteline. Even the first two shape resonances, at about 7758 and 7809 eV, are slightly shifted to lower energies. The absence of a double-structured whiteline, that is typical for hcp and fcc, is at least an indication for ε-Co.

To clarify this, the average core spectrum of Co-nano will be compared to FEFF [75] simulations of the three stable geometrical Co structures: Co-hcp, Co-fcc and Coε (visualization in Fig. 4.2 on page 34; see Table C.3 for crystallographic details). No full multiple scattering (FMS) but a path expansion to $R = 12\,\text{Å}$ is utilized in the FEFF calculations, i.e. Eq. (2.25) instead of Eq. (2.23). Further on, as the Fermi energy is systematically miscalculated by FEFF, a constant energy shift is applied to all simulations. The simulated Co spectra are shown in Fig. 6.13 (right panel) along with the average core spectrum. At first, the fcc-phase can be excluded due to the inappropriate whiteline structure. The ε phase seems to reproduce the structures of the core spectrum best, though its whiteline is slightly too low in energy. Apparently, a mixture of Co-ε and Co-hcp should be most suitable to describe the core spectrum (appropriate instrumental and lifetime broadenings provided).

The average shell spectrum of Co-nano is shown in the left panel of Fig. 6.14, in comparison to its model representative Co-nano-ox. Despite small intensity differences at the edge and whiteline, visualized by the residuum, and an increased width of the shape resonance, shell (avg) nearly resembles Co-nano-ox. This result confirms that a reasonable spectrum has been calculated by the SVD, which was still a matter of discussion due to the bad reproducibility of the CoO contribution of the Mix (see final discussion of last section 6.2.1). To identify the shell spectrum (and Co-nano-ox likewise), it is compared with two common cobalt-oxides CoO and Co_3O_4, measured as powders, shown in Fig. 6.14 (right panel). The shell spectrum differs from CoO, as there is no sharp first shape resonance at about 7740 eV and low intensity of the further shape resonance at about 7770 eV. The positions of these

6. Site-Selective XAS

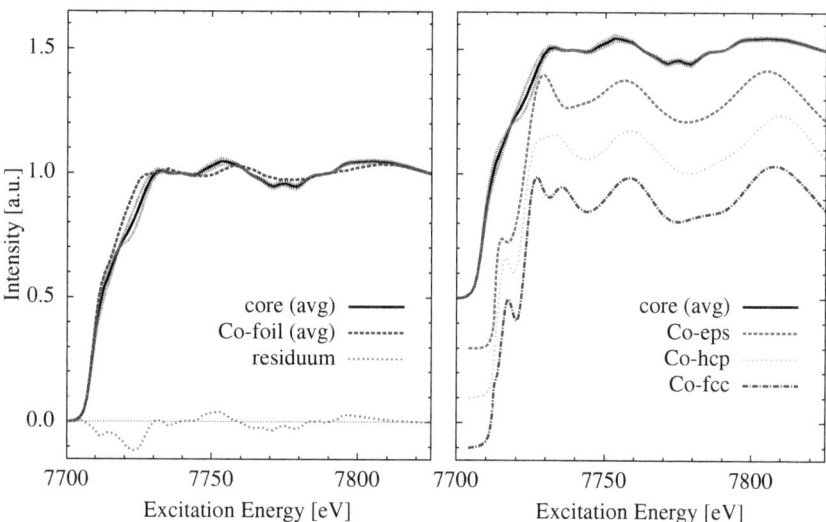

Figure 6.13.: Normalized Co K-edge XANES spectra of the Co nanoparticles' core (with standard deviation) in comparison to the averaged Co-foil (left) and in comparison to FEFF 8.4 simulated ε, hcp and fcc structures of Co (with intensity offset).

shape resonances and of the whiteline are suitable though. The low intensity of the shape resonances is a known effect when dealing with nanoparticles and XANES. It is ascribed to the reduced number of neighboring atoms in a nanoparticle due to its high fraction of surface atoms [56], which is valid for the shell as it consists of a few layers only. However, the absence of the first shape resonance (at 7740 eV) and the width of the second one (at 7770 eV), also points towards the presence of Co_3O_4, though its whiteline is positioned at too high energy due to its higher formal oxidation state of (at least partly) +3. Of note moreover, is the position of the pre-edge feature at about 7708.5 eV for the average shell spectrum. This is in contrast to the standard Co^{2+} reference CoO, where it is shifted about 1 eV to higher energies, but the pre-edge of Co_3O_4 at 7708 eV or a combination of both is more suitable. Based on these findings one could assume the average shell spectrum to be a mixture of Co_3O_4 and CoO. A linear combination fit of the shell spectrum by those two Co oxides gives a ratio of 3:1 (CoO : Co_3O_4), however, with an unacceptable fit quality (R = 168×10^{-5}).

Thus, the origin of the pre-edge feature will be investigated in more detail. For the case of "normal" cubic rocksalt Co(II)O, where Co is octahedrally coordinated by O, quadrupole 1s→3d, dipole 1s→3d/O-2p transitions or even non-local 1s→4p(3d) transitions [19] are possible. Due to the limited resolution of the current measurements, these three types of transitions are responsible for the one pre-edge feature visible in Fig. 6.14. Co_3O_4 in contrast is a spinel, composed of Co(II)O and Co_2(III)O_3, with the former occupying the tetrahedral and the latter the octahedral sites. The Co(II)O part in Co_3O_4 now has the same local atomic environment as wurtzite Co(II)O, and different transitions are possible in both due to the breaking of the inversion symmetry, resulting in an overlap of the Co 3d and 4p bands and consequently a different pre-edge. Actually, it is well known that nanosized CoO in general is not present in the cubic Fm3m phase but mainly in the hexagonal $P6_3mc$ phase

6.2. 1s3p-RIXS at wiggler beamline W1

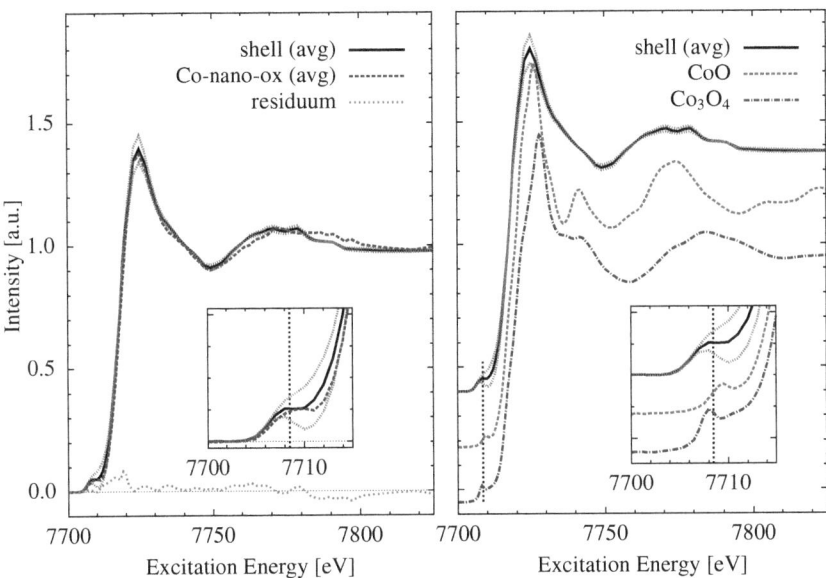

Figure 6.14.: Normalized Co K-edge XANES spectra. Left: The shell (with standard deviation) of the Co nanoparticle in comparison to the averaged Co-nano-ox. Right: The shell of the Co nanoparticle as well as a cobalt(II)-oxide (CoO) and a cobalt(II,III)-oxide (Co_3O_4). The latter two spectra are offset with respect to the intensity. The vertical line at 7708.5 eV indicates the position of the pre-edge feature of the average shell spectrum. The insets show a magnification of the pre-edge regions.

[112, 82, 59]. Thus, it is likely that the significant differences between CoO and Co_3O_4, visible in the pre-edge in Fig. 6.14, can be attributed to wurtzite-CoO.

Noteworthy is that another reason for the pre-edge shift to lower energies could be a less electronegative partner, i.e. carbon instead of oxygen. Since the presence of C is possible due to the precursor $Co_2(CO)_8$ and the reactant $Al(C_2H_5)_3$, this is also a possible explanation of the observed shift.

All in all, the extraction of two pure site-selective XANES spectra from HRFD-XANES by utilizing a SVD was successful. For the cobalt nanoparticles under investigation site-specific spectra have been extracted for the two main sites: a metallic Co core with a crystal structure that is most likely a mixture of Co-ε and Co-hcp and a CoO shell of valency two, exhibiting the cubic as well as the hexagonal phase, with a mixture of O and most probably C as ligands.

6.3. 1s3p-RIXS, HRFD-XANES and VTC-XES at undulator beamline ID26

The complete section is published in [50].

6.3.1. Experimental

The undulator beamline ID26 at the 6.0 GeV storage ring of the ESRF provides a flux of more than 10^{13} photons/s (see section 3.4 and Table 3.2 for more details). The incident energy was tuned through the Co K-edge by means of a Si(111) cryogenically-cooled fixed-exit double-crystal monochromator. The subsequently emitted Kβ radiation was detected by an emission spectrometer that uses four spherically bent Ge(444) analyzer crystals (Ge(111) utilized in fourth order). The experimental resolution of both devices was about 1 eV. The following measurements were performed on cobalt nanoparticles with increasing shell to core ratio from Co-nano-1 to Co-nano-2, to Co-nano-3 (see section 4.1.1), as well as on references metallic Co foil, cobalt(II)-oxide (CoO) and cobalt(II)-carbonate ($CoCO_3$):

1. Non-resonant emission scans (NRXES) of the complete Kβ emission line ($K\beta_{1,3}$ and $K\beta_{2,5}$) of all six samples, at fixed excitation energy of 7800 eV.

2. 1s3p RIXS measurements of the three references Co-foil, CoO and $CoCO_3$, gained by successively measuring HRFD-XANES spectra (described in the next point), while scanning the $K\beta_{1,3}$ fluorescence.

3. HRFD-XANES and EXAFS measurements of the three references Co-foil, CoO and $CoCO_3$, gained by detection of fluorescence at the $K\beta_{1,3}$ peak positions of each reference, while tuning the incident energy.

4. HRFD-XANES and EXAFS measurements of the three Co nanoparticles from four different positions (see Table 6.5) on the $K\beta_{1,3}$ emission line, gained by consecutive detection of fluorescence at these positions.

5. HRFD-XANES measurements of the three Co nanoparticles from two different positions (see Table 6.5) on the $K\beta_{2,5}$ emission line, gained by consecutive detection of fluorescence at these positions.

The stepwidths and collection times of each type of measurement are listed in Table 6.5. For the EXAFS scans the samples have been cooled additionally by means of a displex cryostat with the sample in vacuum and using He as cryogen.

6.3.2. Overview

The complete NRXES Kβ emission up to the Fermi energy, including $K\beta_{1,3}$ and $K\beta_{2,5}$, of the references Co, CoO and $CoCO_3$ is shown in Fig. 6.15. The spectra have been normalized with respect to the spectral area and are aligned according to the first moment (area centroid). At low energies (about 7637 eV and 7649 eV) the strong Kβ' and $K\beta_{1,3}$ lines are visible, which reflect the 3p→1s transitions (splitted by the 3p-3d exchange interaction, see section 2.3.2). At high energies (about 7685 eV and 7706 eV) the weak satellite emission lines Kβ'' and $K\beta_{2,5}$ are just barely visible. In the inset thus, they are magnified by about a factor of 35, and their complex structure, attributed to the various ligand (hybridized with the metal valence orbitals) to 1s core transitions, is recognizable.

6.3. 1s3p-RIXS, HRFD-XANES and VTC-XES at undulator beamline ID26

Table 6.5.: Details about the various measurements at beamline ID26 described in section 6.3.1.

description		Range [eV]	steps	stepwidth [eV]	collection time per spectrum [s]
1. NRXES	$K\beta_{1,3}$	7620 - 7662	210	0.2	210
	$K\beta_{2,5}$	7660 - 7725	163	0.4	1630
2. RIXS	excitation (1s)	7650 - 7800	1109	0.1	60
	emission (3p)	7630 - 7656	130	0.2	
3. - 5. HRFD-	XANES	7690 - 7800	1105	0.1	120, 240^a
	EXAFS	7489 - 8483	1109	1.0	300
		Position [eV]			
3. HRFD-XANES/EXAFS		7649.6 ($K\beta_{1,3}$ peak of Co)			
		7650.8 ($K\beta_{1,3}$ peak of CoO)			
		7650.9 ($K\beta_{1,3}$ peak of CoCO$_3$)			
4. HRFD-XANES/EXAFS		Pos-0: 7636.6 ($\sim K\beta'$ peak of CoO)			
		Pos-1: 7644.4 (low-energy side of $K\beta_{1,3}$)			
		Pos-2: 7649.6 ($K\beta_{1,3}$ peak of Co)			
		Pos-3: 7652.0 (high-energy side of $K\beta_{1,3}$))			
5. HRFD-XANES		Pos-4: 7702.6 ($K\beta_{2,5}$ peak of CoO)			
		Pos-5: 7706.6 ($K\beta_{2,5}$ peak of Co)			

a nanoparticles with longer collection time than references

The HRFD-XANES spectra of the three references, extracted at their respective $K\beta_{1,3}$ peak positions (see point "3." in Table 6.5), are shown in Fig. 6.16 in the left panel. The high resolution is immediately apparent, as for e.g. Co-foil a clear double-whiteline is visible for the first time that reflects its hcp-fcc mixture in accordance to simulations (see appendix C). The superior resolution is furthermore demonstrated by comparing the HRFD-XANES spectrum of, e.g., CoO with its total-fluorescence as well as its classical transmission spectrum in the right panel of Fig 6.16. Only by measuring a single fluorescence channel, as is done in HRFD-XANES, the pre-edge can be clearly resolved. As to the rest, the CoO spectral features of the total-fluorescence scan are already far better resolved than in the transmission scan (measured at ANKA's INE beamline, see section 3.2). Furthermore, the shifted edge of CoO to higher energies ($\Delta E \simeq 7$ eV) relative to Co-foil, is visible clearly for the first time, and the almost unstructured whiteline of CoO in transmission reveals itself in fluorescence to be made of one peak with two clear shoulders at lower and higher energies.

This superior resolution will be exploited upon performing the following tasks, which can be related to the complete $K\beta$ spectrum in Fig. 6.15 as follows:

- Investigation of HRFD-XANES as well as HRFD-EXAFS spectra recorded at appropriate positions at the main $K\beta_{1,3}$ emission line, to achieve valency- and site-selectivity by means of the general strategy elaborated in section 6.1.

- Investigation of the satellite emission line $K\beta_{2,5}$ (and $K\beta''$), to identify the types and conditions of the ligands. Furthermore, the HRFD-XANES spectra recorded at $K\beta_{2,5}$ positions will be

6. Site-Selective XAS

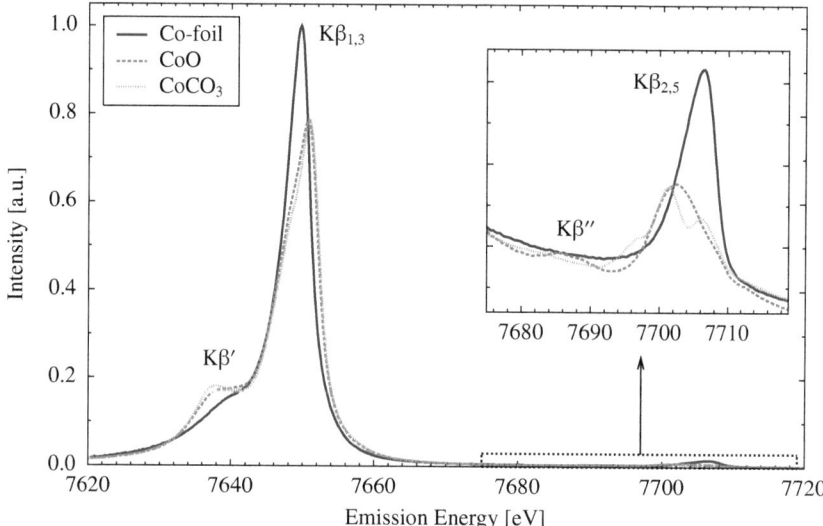

Figure 6.15.: Normalized Kβ emission spectra of the three references Co, CoO and CoCO$_3$. The main Kβ$_{1,3}$ line and Kβ', as well as the ligand sensitive Kβ$_{2,5}$ line and crossover resonance Kβ" are shown. The inset shows a zoom to the latter, with ×35 magnification.

studied, to check for site-selectivity with respect to the ligand-sites.

6.3.3. NRXES Kβ$_{1,3}$ spectra

The normalized (with respect to the spectral area) and aligned (with respect to the areas centroid) NRXES Kβ$_{1,3}$ spectra of the three Co nanoparticles and the three references are shown in the two panels of Fig. 6.17. The chemical shift between zerovalent Co-foil and divalent CoO is about 1.1 eV, identical to previous results. The Kβ$_{1,3}$ peak of CoCO$_3$ is slightly shifted to higher energies (about 0.1 eV) and the Kβ' peak to lower energies compared to CoO, indicating a slightly higher net valence spin of CoCO$_3$, albeit it is formally divalent like CoO. The three Co nanoparticles' Kβ$_{1,3}$ spectra are shown in the right panel of Fig. 6.17 and do show only slight differences. Furthermore, their peak positions are almost identical to Co-foil as can be recognized by the first arrow. However, from the synthesis (see section 4.1.1) an increase of the Co^{2+} fraction from Co-nano-1 to 3 is expected, and a closer look to both peak regions (with the help of the zooms) confirms this due to the following effects: intensity gain of the Kβ' peak and intensity loss and slight shift to higher energies of the Kβ$_{1,3}$ peak from Co-nano-1 to 3.

To determine the zero- and divalent Co contribution more quantitatively, linear combination fits (LCF's) of the Kβ$_{1,3}$ spectra of all three nanoparticles by Co-foil and CoO are performed. Actually, CoCO$_3$ is included in these LCF's too, but without significant contribution. The results are given in Table 6.6 and Fig. 6.18 and show satisfying fit qualities, so that Co-foil and CoO are taken as model compounds from now on. The fit spectra are shown in the first three panels of Fig. 6.18, whereby

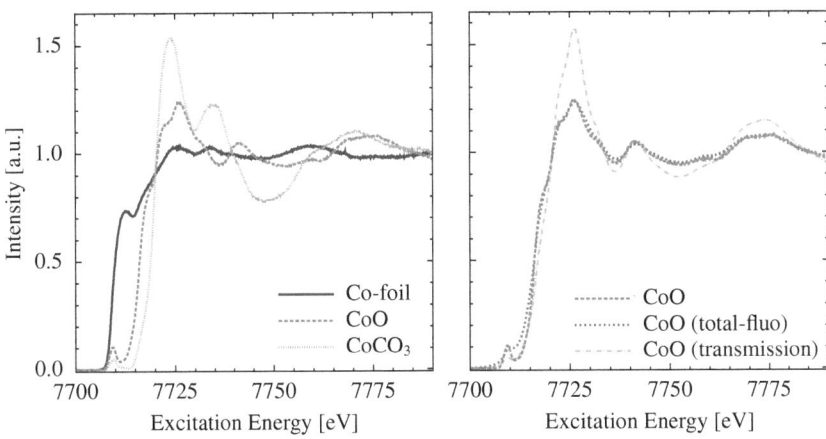

Figure 6.16.: Normalized Co K-edge HRFD-XANES spectra of (left) Co, CoO and CoCO$_3$, extracted at their K$\beta_{1,3}$ peak positions (see point 3. in Table 6.5), and (right) the same CoO spectrum in comparison to its total fluorescence and transmission scan.

Table 6.6.: Overall Co-foil to CoO ratios as obtained via the LCF of the K$\beta_{1,3}$ spectra of Co-nano-1, 2 and 3, as well as for the specific positions Pos-1, 2 and 3. Below are the corresponding ratios from the LCF of the HRFD-XANES spectra. The R-factor describes the fit quality. The fit errors (given in brackets) of Pos-1, 2 and 3 are similar to those from the overall fits, correlated to the R-factor though.

LCF K$\beta_{1,3}$	R-factor [e-5]	overall ratio	Pos-1	Pos-2	Pos-3
Co-nano-1	64	80.5(4) : 19.5(4)	80.3 : 19.2	87.4 : 14.9	71.9 : 28.3
Co-nano-2	32	72.6(3) : 27.4(3)	72.5 : 27.0	80.1 : 21.3	62.1 : 38.1
Co-nano-3	8	58.5(1) : 41.5(1)	59.3 : 41.7	67.2 : 33.7	46.5 : 53.8
LCF XANES	(overall; Pos-1, 2, 3)				
Co-nano-1	19; 17, 9, 40	86.7(3) : 13.3(3)	77.8 : 22.2	89.8 : 10.2	66.3 : 33.7
Co-nano-2	33; 48, 47, 86	63.1(4) : 36.9(4)	62.1 : 37.9	65.8 : 34.2	49.2 : 50.8
Co-nano-3	19; 40, 42, 116	61.4(3) : 38.6(3)	55.0 : 45.0	60.8 : 39.2	38.2 : 61.8

the intensity of the spectra of the model compounds are already weighted with respect to the fitting results. The overall ratios given in Table 6.6 are proving the Co^{2+} increase from Co-nano-1 to 3. According to these ratios one can estimate the sizes of the Co core and CoO shell (see appendix A and compare section 6.2.2), on the assumption that of spherically shaped nanoparticles and by using the bulk values for the density of Co and CoO: Co-nano-1, 2 and 3 with core diameter about 5.3 nm, 5.1 nm and 4.6 nm and shell thickness about 0.35 nm, 0.45 nm and 0.70 nm. Thus, Co-nano-3 exhibits the thickest shell with about 2 - 3 layers of CoO (equal to the nanoparticles in section 6.2.2) and Co-nano-1 the thinnest with about a monolayer of CoO.

6. Site-Selective XAS

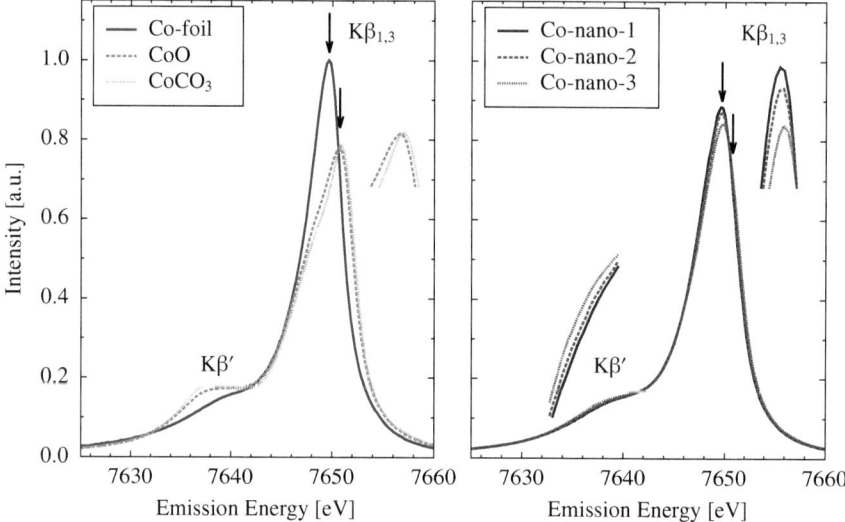

Figure 6.17.: Normalized NRXES K$\beta_{1,3}$ spectra of the three references (left) and the three nanoparticles (right). The two arrows indicate the peak position of Co-foil and CoO, respectively. For the nanoparticles' spectra both peaks are also shown magnified.

6.3.4. Valency/Site-selective XANES

In the last panel of Fig. 6.18 the fractions of Co-foil and CoO with respect to the nanoparticles are shown. With the help of this figure the positions (indicated by arrows and vertical lines respectively) with intermediate (Pos-1), lowest (Pos-2) and highest (Pos-3) CoO ratio are defined (see Table 6.5 for precise values of positions). The explicit ratios at these positions are given in Table 6.6 too, and obviously Pos-1 almost resembles the overall ratio while it is increased and decreased about 10 % for Pos-2 and Pos-3, respectively. At these positions HRFD-XANES spectra of the nanoparticles have been measured. They also have been fitted by the HRFD-XANES spectra of the reference compounds Co-foil and CoO (results in Table 6.6), measured at identical positions. The nanoparticles' HRFD-XANES spectra are shown in Fig. 6.19 in two alternative representations: In the left panel the spectra are sorted with respect to the nanoparticles (1, 2, 3) and in the left panel with respect to the positions (1, 2 ,3). In both representations the partial site-selectivity to metallic Co or divalent CoO (or CoCO$_3$) is clearly recognizable in the variation of the edge and the whiteline (compare Fig. 5.2 in section 5.1 and explanations given there). The reason for the two representations is that there are two possible "methods" for the extraction of site-selective spectra at hand, which can be schematized in the following way (each "+" correlates to one spectrum and (a) to (f) refer to the sub-figures of Fig. 6.19):

6.3. 1s3p-RIXS, HRFD-XANES and VTC-XES at undulator beamline ID26

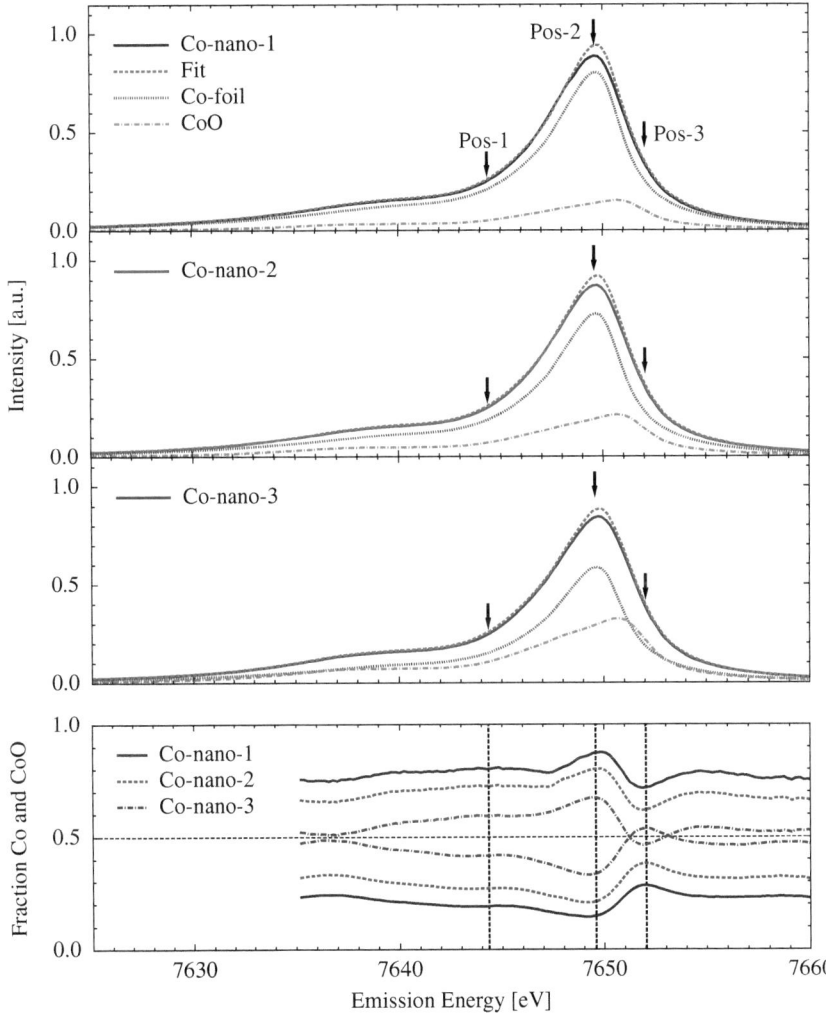

Figure 6.18.: First three panels: LCF's of the three $K\beta_{1,3}$ NRXES spectra of the Co nanoparticles by the references Co and CoO. The bottom panel shows the fractions of Co (first three curves) and CoO (last three curves) in Co-nano-1, 2 and 3 as determined by the LCF's. The three arrows and the vertical dashed lines, respectively, indicate the positions from which the HRFD-XAS spectra are extracted.

6. Site-Selective XAS

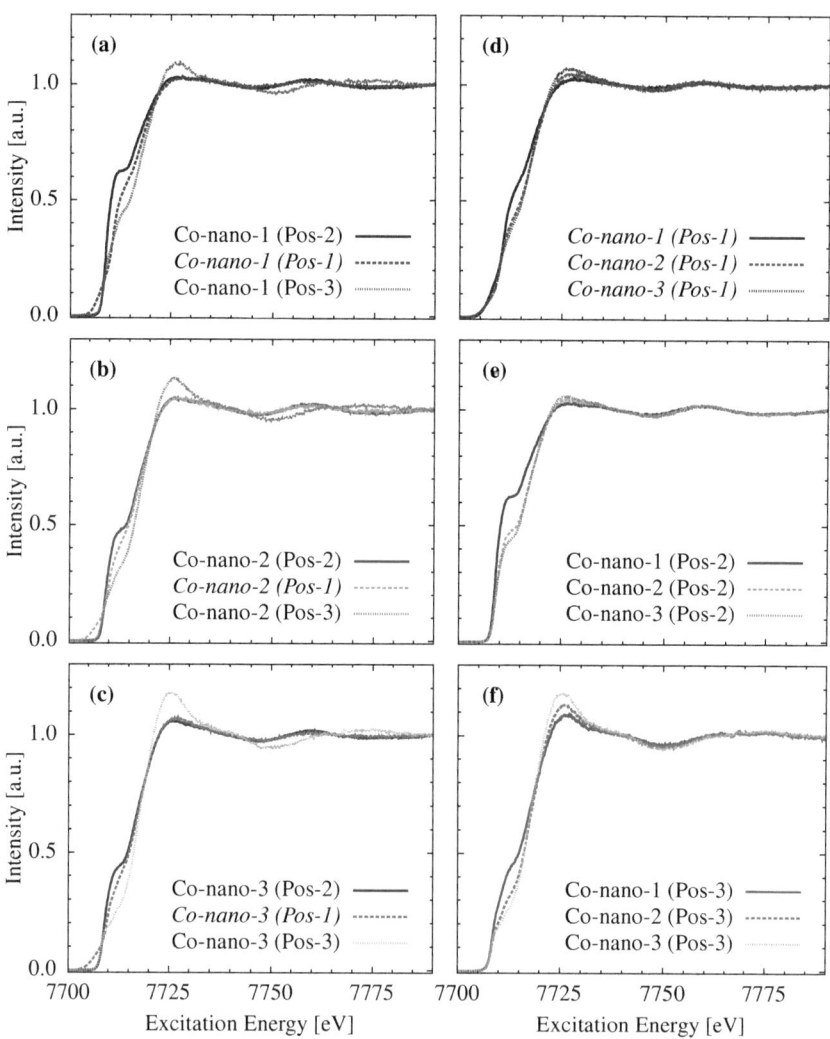

Figure 6.19.: Normalized Co K-edge HRFD-XANES spectra of the Co nanoparticles. Left panel (Method 1): Comparison of spectra from Pos-2, 1 and 3 of Co-nano-1 (a), 2 (b) and 3 (c) respectively. Right panel (Method 2): Comparison of spectra of Co-nano-1, 2 and 3 measured at Pos-1 (d), 2 (e) and 3 (f). The spectra with labels written in cursive are excluded in the further process as explained in the text.

6.3. 1s3p-RIXS, HRFD-XANES and VTC-XES at undulator beamline ID26

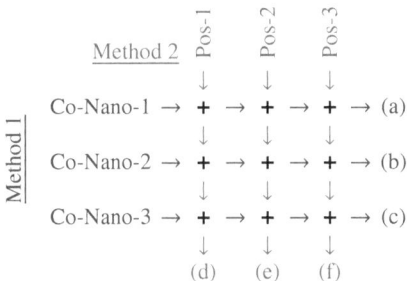

Method 1 (blue horizontal arrows) is identical to the one applied in the last section (6.2.2 Co nanoparticles), i.e. the variation of the core to shell ratio is achieved by making use of the partial site-selectivity of HRFD-XAS spectra (by extracting them at different positions Pos-1, 2, 3). Here from each set of HRFD-XANES spectra (a), (b), and (c) of Fig. 6.19, site-selective spectra will be determined. Thus, three pairs of core and shell spectra will be obtained as there are three (slightly different) Co nanoparticles, which can be compared then with one another. Two possible solutions are conceivable: (a) Due to the shell's bondage to the core, there could be some interaction, which might lead to variations of core properties in dependence of the shell's thickness. (b) If this interaction is rather weak, the core's properties will be identical for each of the three nanoparticle types. Finally, as in the last section 6.2.2, care has to be taken due to the lifetime influences onto these HRFD-XANES spectra (also see section 2.3.3), what can (and will) be checked easily.

In Method 2 (magenta vertical arrows) the core to shell ratio is varied due to three different syntheses of the Co nanoparticles (see section 4.1.1). Thus, for each of the three positions (Pos-1, 2, 3) site-selective spectra will be determined from each set of HRFD-XANES spectra (d), (e) and (f), respectively, of Fig. 6.19, without disturbance due to the lifetime. To be precise the lifetime broadenings do influence the spectra, but for one position the influence is equal for each of the three (Co-nano-1 to 3) HRFD-XANES spectra. However, if one compares the three (for each fluorescence position) independently determined core and shell spectra, the lifetime influences have to be taken into account again.

It should be mentioned here that upon comparison of the final averaged results of both methods the total lifetime disturbances are equal in each case. Furthermore, a third method is possible here, by fitting all HRFD-XANES spectra simultaneously (method S). This method should be the most unsusceptible towards non-uniform distributions of the components Co and CoO in the small sets of HRFD-XANES spectra used in method 1 and 2, that could result in under- or overestimations of one of the site-specific components (what will be understandable in the following). Before going on, LCF's of all HRFD-XANES spectra of Fig. 6.19 by the model compounds, taken at the same positions Pos-1, 2 and 3 and also from the total fluorescence yield (overall), are performed. The results are given in Table 6.6 too, just below the $K\beta_{1,3}$ LCF and show an increased sensitivity with respect to CoO about 5 % to 10 %. For the Co-CoO test system (section 6.2.1) the $K\beta_{1,3}$ LCF ratios have been the most suitable to get the true site-specific spectra. Thus, if possible, the SVD ratios will be adjusted to those for the current study too.

Proceeding with the strategy to get site-selective spectra (section 6.1), it turned out that Method 1 does not work unless the spectra from Pos-1 are excluded. Upon reviewing the spectra (a) - (c) in Fig. 6.19 it is obvious that those Pos-1 spectra exhibit a significantly different edge onset in comparison to those from Pos-2 and 3. Remembering Fig. 6.18 where the positions were marked by arrows, one can see that Pos-1 has the lowest intensity. Furthermore, it is very close to the $K\beta'$ peak. Thus, Pos-1 is influenced by lifetime broadenings (recall section 2.3.3) from resonances of both $K\beta'$ and

6. Site-Selective XAS

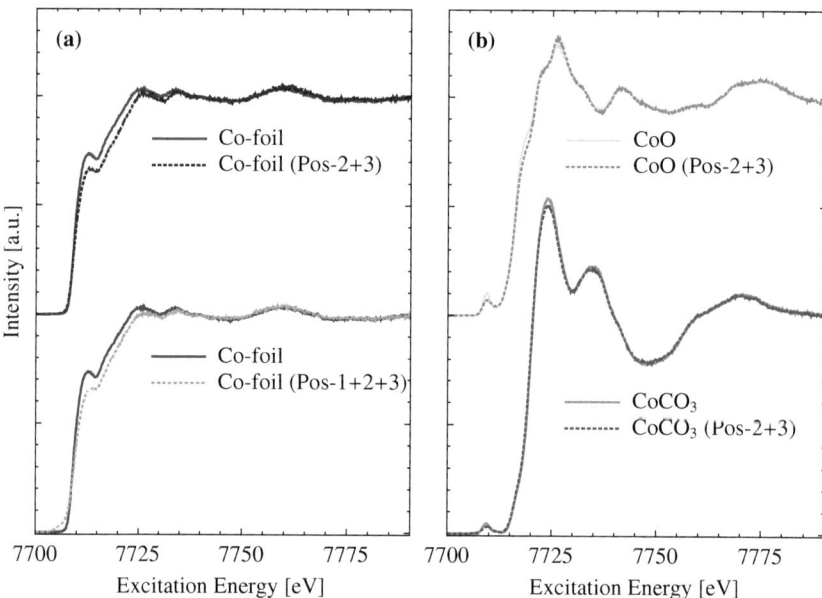

Figure 6.20.: Normalized HRFD-XANES spectra of the three references Co, CoO and CoCO$_3$ in comparison to their respective averaged spectra from Pos-2 and 3 (and also from Pos-1, 2 and 3 for Co-foil).

K$\beta_{1,3}$, and due to the low intensity at Pos-1, these influences show up much stronger than at Pos-2 and 3. Thus, by excluding the Pos-1 spectra, no physical information is neglected. The different edge is just a matter of lifetime broadenings which, however, could have been circumvented by choosing a fluorescence position more close to the main K$\beta_{1,3}$ peak, i.e. at higher intensities comparable to Pos-3.

At this point the suitability of the two positions left will be checked. Therefore, for each of the three references Co, CoO and CoCO$_3$, the average of the HRFD-XANES spectra from Pos-2 and 3 is compared to the respective HRFD-XANES spectrum recorded at the fluorescence peak in Fig. 6.20. As was elucidated in section 2.3.3 the average spectra should as best resemble those recorded at the peaks, to assure the validity of the to be determined site-selective spectra. For the divalent model compounds CoO and CoCO$_3$ the two positions 2 and 3 are obviously most suitable as only slight intensity differences show up. For the zerovalent Co-foil, however, it is worse: All features are equally present, but for Co-foil (Pos-2+3) the edge feature at 7713 eV and the first whiteline peak at about 7725 eV are lower in energy. Furthermore, both its whiteline peaks and the shape resonance at about 7760 eV are shifted about +0.5 eV. Even upon inclusion of the Pos-1 spectrum, these differences show up as can be seen in the lower part of Fig. 6.20 (a). Moreover, the edge onset gets significantly worse due to the Pos-1 spectrum, which is why it can not further be used in the SVD. The reason why the averaged spectra of CoO and CoCO$_3$ match those from the peaks but Co-foil not, is of course that the K$\beta_{1,3}$ peaks of CoO and CoCO$_3$ are just positioned in the center between Pos-2 and 3, while that

6.3. 1s3p-RIXS, HRFD-XANES and VTC-XES at undulator beamline ID26

Table 6.7.: Ratios of core : shell as obtained by the SVD of the Co nanoparticles' HRFD-XANES spectra from Pos-2 and 3. Results of three different methods (S, 1 and 2) are shown. The triplet of R-factors gives the fit quality for Co-nano-1, 2 and 3. The corresponding fit errors are always ±0.2 except for fits with R > 30 where it is ±0.4.

	method S		method 1 ($\rightarrow$)		method 2 ($\downarrow$)	
	Pos-2	Pos-3	Pos-2	Pos-3	Pos-2	Pos-3
Co-nano-1	98.4 : 1.6	62.3 : 37.7	100.0 : 0.0	66.0 : 34.0	100.0 : 0.0	67.0 : 33.0
Co-nano-2	74.4 : 25.6	43.7 : 56.3	80.2 : 19.8	46.3 : 53.7	78.2 : 21.8	45.3 : 45.7
Co-nano-3	69.2 : 30.8	34.7 : 65.3	74.5 : 25.5	36.2 : 63.8	72.5 : 27.5	35.9 : 64.1
R-factor [e-5]	11, 13, 9	13, 10, 8	39, 5, 5	10, 8, 5	33, 12, 7	9, 8, 38

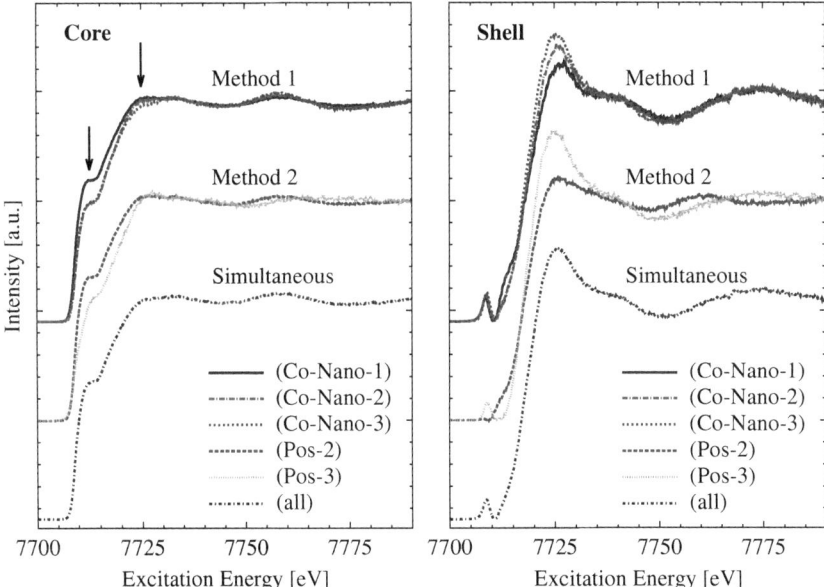

Figure 6.21.: Site-specific Co K-edge spectra of the Co nanoparticles core (left) and shell (right) respectively. The spectra are calculated from HRFD-XANES spectra as given in brackets in the labels and corresponding to the methods 1, 2 and S.

of Co-foil is equal to Pos-2 (see Table 6.5), and thus the Pos-3 spectrum causes the differences. As there are no other HRFD-XANES spectra (from different positions) available, the differences visible for Co-foil have to be kept in mind, when it comes to the interpretation of the site-selective spectra.

Now method 1 and consequently also method 2 and S are performed without the Co nanoparticles' spectra from Pos-1. The least squares fit upon employment of the SVD (see last section 6.2.2 or appendix B and Ref. [84]) has to be applied three times with 2 spectra for method 1 and two times

6. Site-Selective XAS

with three spectra for method 2 and one times with six spectra for the third method. In contrast to the previously performed SVD, it turned out that upon fitting the SVD ratios to those of the LCFs (Table 6.6), the "physical" boundaries of the SVD (ratio between 0 and 1 and intensity ≥ 0) always restricted the results to one unique solution. In other words, only one pair of spectra for core and shell is obtained in order to get as close as possible to the LCF ratios, which connotes an improvement to the previous study (section 6.2.2). The reason for this can be addressed to the higher resolution of the HRFD-XANES spectra measured at the ESRF in contrast to those from HASYLAB, which resulted in less unstructured spectra and hence less variability in the SVD. The resulting core and shell spectra are shown in Fig. 6.21 and the corresponding ratios are given in Table 6.7. Obviously, neither all ratios from $K\beta_{1,3}$-LCF nor from XANES-LCF could be reproduced, albeit some single results are almost identical, but the overall trend was preserved. Upon confronting all results from all three methods (shown in Fig. 6.21) the following can be stated:

- Method 1: Obviously all three spectra for core and shell, respectively, are quite similar with respect to energetic positions of the features. Intensity differences are visible though, in particular at the edge for both the core and shell spectrum calculated from Co-nano-1 (dark-blue lines). One could argue that the intensity differences of the core spectra are due to the influence of the different shell thicknesses. Then the most intense whiteline (indicated by second arrow in Fig. 6.21 left), visible for the core of Nano-1, can only be explained by a higher contribution of e.g. the Co-fcc phase to standard Co-hcp (vide infra Fig. 6.23). This, however, contradicts with the intensity differences at the edge (first arrow), that imply differences in the degree of metallicity. Furthermore, it will be shown later by EXAFS analysis (section 6.3.5 on page 94) that the Co-hcp dominates all three Co-nanoparticles likewise. Consequently, the differences must be attributed to faults in the numerical procedure (SVD). Co-nano-1 for example has the smallest shell fraction (see Tables 6.6 and 6.7), which makes it poorly representative regarding the shell and which leads to an overestimation of the metallic core part. The same can be inferred for the differences visible at the three shell spectra, since although the intensity decrease and energy increase of the whiteline's suggest an increase of the valency (from Nano-3 to 1), the alikeness of pre-edge and edge for all three nanoparticles contradicts this assumption (vide infra Fig. 6.24 b), so that likewise for the core, an underestimation of the shell part can be concluded for Co-Nano-1. As oppositely, there could be an underestimation of the core and overestimation of the shell for Co-nano-3, the three core and shell spectra, respectively, of method 1 will be averaged, so that the contradicting effects are cancelled out. All in all, method 1 yields unique core and shell XANES spectra for all three nanoparticles, albeit distorted owing to the strongly differing representativity of core and shell by the singular nanoparticles.

- Method 2: Here by contrast, the two spectra for core and shell are significantly different. Both show strong differences in the edge and whiteline region and furthermore even in the shape resonance above 7750 eV, which position differs by about 15 eV for the two core and two shell spectra. Actually the core spectrum from Pos-3 and the shell spectrum from Pos-2 are the ones that are significantly different from the complete results of method 1. The reason again is the bad representativity of the former regarding the core and the latter regarding the shell (compare ratios in Tables 6.6 and 6.7). Since the deviations here are more drastic compared to method 1 and since each site-specific spectrum represents the average of the three nanoparticles and so actually must be equal for core and shell respectively, the results from this method will be discarded.

- The simultaneous SVD in Fig. 6.21 eventually yields two site-specific spectra that fit well

6.3. 1s3p-RIXS, HRFD-XANES and VTC-XES at undulator beamline ID26

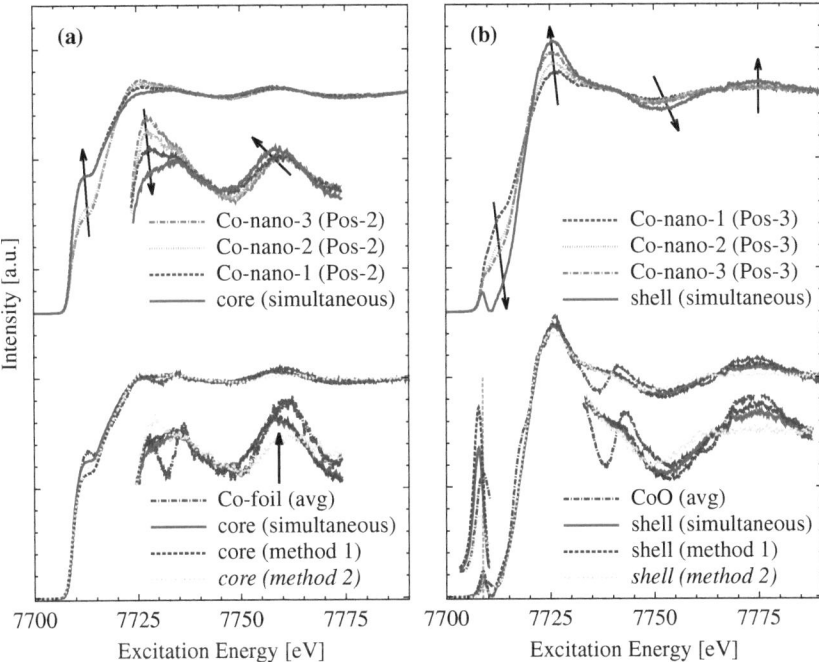

Figure 6.22.: Left (Right): Site-specific XANES spectrum of the Co nanoparticles' core (shell) as calculated by SVD (method given in brackets) along with its model compound Co-foil (CoO) at the bottom and along with the HRFD-XANES spectra of the Co nanoparticles from Pos-2 (Pos-3) at the top.

to the average spectra of method 1, corroborating its results as well as the discard of those from method 2. Furthermore, it gives the importance of all six (equal to number of experimental spectra) possible site-specific components: 95.23 %, 3.40 %, 0.69 %, 0.48 %, 0.12 % and 0.08 %. These values should not be mixed up with the components' ratios. They just give the importance with respect to the SVD. Hence, upon inclusion of two components (the core and the shell) the fit is almost perfect and can just be improved by less than 2 %, if further components are to be incorporated.

For the further interpretation of the site-specific results, the core and shell spectra of method 1 (and 2) will be averaged and then compared to those of method S, as well as to the model compounds, in the lower part of Fig. 6.22. The spectra of method 2 are shown too for completeness and to demonstrate the failure of method 2, even upon averaging of the single results. The spectra of method S and 1 do nearly coincide for both core and shell, though the core's edge at 7712.5 eV of method 1 is weaker. Furthermore, the core spectra resemble the Co-foil spectrum, however, the whiteline lacks the double peaks and the first shape resonance is shifted to lower energies about 1.5 eV to 7758.0 eV. The shell spectra are similar to the one obtained in the previous section 6.2.2 (see Fig. 6.14), i.e. compared to CoO: pre-edge slightly shifted to lower energies (7708.7 eV), first shape resonance (at about 7741 eV)

6. Site-Selective XAS

nearly absent and second shape resonance (around 7775 eV) broader. The latter two effects can be attributed to the small size of the nanoparticles.

The core and shell spectrum from the simultaneous method will be kept for further investigation. This is justified by their comparison to the Co–rich and CoO–rich Co-nano spectra in the upper parts of Fig. 6.22. The orientation of the arrows is towards metallic Co enhancement in the left panel and towards divalent CoO enhancement in the right panel. Obviously the core and the shell spectra are in accordance to the trends indicated by the arrows, just the core's edge should be stronger, not weaker, and hence the preference of the result from method S versus method 1.

Analysis of site-specific XANES spectra

In Fig. 6.23 the core spectrum, whose features are marked by dashed vertical lines, is compared to simulations of the known metallic Co-phases hcp, fcc, ε and bcc (see Fig. 4.2 for visualization and Table C.3 for crystallographic details) and the reference, Co-foil. Two sets of simulations have been performed which both are based on multiple scattering (MS) theory (see section 2.2.2), however, upon utilizing different exchange correlation potentials. In the left panel of Fig. 6.23 the FEFF9 [75] code has been applied and in the right the FDMNES [42] code (see appendix C for computational details). Actually all features are identically calculated by both codes, except for the splittings which are smaller for FDMNES being more appropriate to describe edge and whiteline separation and larger for FEFF, suitable for shape and position of whiteline and first shape resonance. This difference in the splitting of the features is a general effect occurring from the different potentials applied and will be encountered in the other simulations of Co compounds too. Furthermore, for the Co-hcp simulation by FDMNES the whiteline at around 7726 eV shows a decreasing slope in contradiction to the literature and FEFF tests (see appendix C). Consequently, for the interpretation of Co metal and the nanoparticles' core, the FEFF simulations will be consulted. As the energetic positions of the simulated features are partially incorrect, all interpretations can only be made upon comparing the different phases among themselves.

- The most suitable candidate for the core spectrum is Co-hcp which also is shown magnified at the bottom of Fig. 6.23 (a). It can be inferred however, that a second component is necessary as the slope of the hcp whiteline is slightly to steep. Another hint for a second contributing phase is the shift of the core's first shape resonance to lower energies about 1.5 eV – remember the magnification at the bottom of Fig. 6.22 (a) – which can not be attributed to differences caused by lifetime-influences as shown in Fig 6.20 (the shift was negative there). As the position of the first shape resonance of Co-hcp is identical to Co-fcc and Co-foil, a slight admixture of the Co-ε or Co-bcc phase is required.

- Co-ε as the main phase, assumed for the core in section 6.2.2 and Fig. 6.13, can not be confirmed hence. However, one of the extreme cases for that previous core spectrum (see Fig. 6.12) at least resembled a Co-hcp phase, though with low resolution. One can conclude thus that the lower resolution of those earlier measurements led to low-resolution site-specific spectra which were hard to identify correctly.

The shell spectrum is compared to the spectra of the model compounds CoO-avg and $CoCO_3$-avg (both averaged from Pos-2 and Pos-3) in Fig. 6.24 (a). Obviously the shell spectrum does not resemble one of the model compounds solely, but the LCF by both (shown in the same figure) gives an acceptable reproduction of the main features with ratio CoO : $CoCO_3$ = 78.1 : 21.9 (error ±0.8) and fit quality of $R = 68 \times 10^{-5}$. However, neither the pre-edge of the shell spectrum at 7708.7 eV could be

6.3. 1s3p-RIXS, HRFD-XANES and VTC-XES at undulator beamline ID26

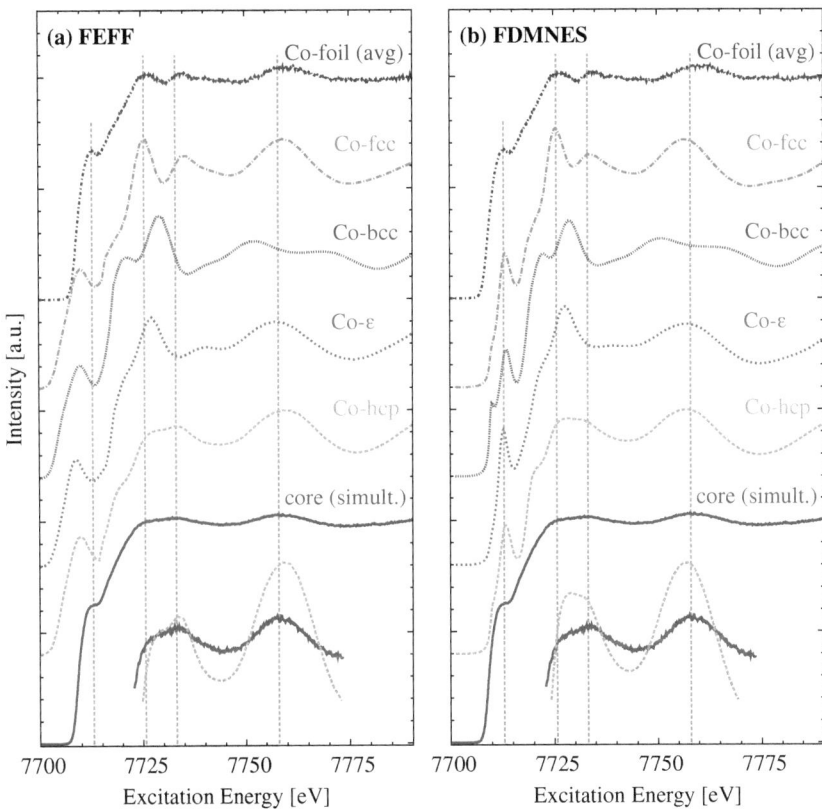

Figure 6.23.: The site-specific XANES spectrum of the Co nanoparticles' core in comparison to normalized Co K-edge XANES spectra of the metal Co phases, simulated by FEFF9 (a) and FDMNES (b) and the Co-foil. The vertical dashed lines mark the significant features of the core spectrum. A magnification of a part of the core spectrum along with Co-hcp is given at the bottom.

fitted, due to its shift to lower energies, nor the shape of its structureless whiteline at about 7726 eV, albeit its energetic position is correctly reproduced. Furthermore, the first two shape resonances at about 7741 eV and 7774 eV, respectively, are less pronounced though attributable to a size-effect of the nanoparticle's shell. Consequently, the presence of more and/or different Co compounds in the shell is most probable. In Fig. 6.24 (b) the shell spectrum is compared to the different stable Co-O compounds, measured in transmission at the INE beamline at ANKA (see section 3.2). On the one hand the formal oxidation number of (mainly) +2 can be verified, on the other hand an explanation for the low-energetic pre-edge arises due to Co_3O_4 ($Co^{II,III}O$) or even Co_2O_3 ($Co^{III}O$). If truly a $Co^{III}O$ compound contributes, its fraction has to be small, since the whitelines of Co_3O_4 and Co_2O_3 at 7728 eV and 7729 eV, respectively, are too high in energy regarding the shell spectrum.

6. Site-Selective XAS

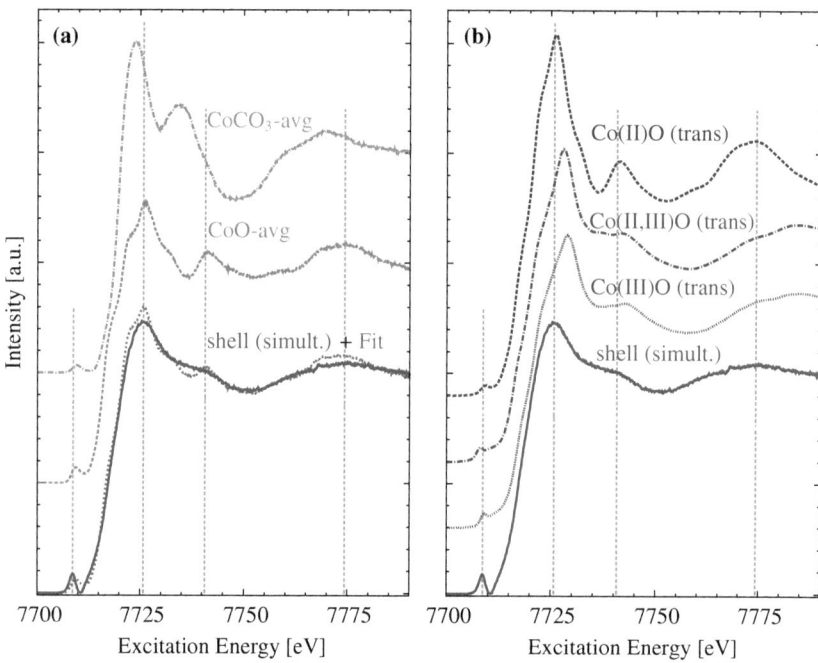

Figure 6.24.: The site-specific XANES spectrum of the Co nanoparticles' shell in comparison to normalized Co K-edge XANES spectra of Co-O/C references, measured at the same beamline (a) and elsewhere in transmission (b). Significant features of the shell spectrum are marked by dashed vertical lines.

To clarify the origin of the low-energetic pre-edge and to continue the respective discussion at the end of section 6.2.2, respectively – where wurtzite type CoO was suggested as the most probable candidate for the shell (see also Fig. 6.14) – simulations of CoO in the rocksalt (CoO-cub) and the wurtzite phase (CoO-hex) as well as Co_3O_4 in the diamond phase (see Table C.3 for crystallographic details), are shown in Fig. 6.25. The gray dashed vertical lines mark the significant features of the experimental spectra CoO and Co_3O_4 (measured in transmission) that are also shown and to which the two sets of calculations have been aligned (see appendix C). The main findings are:

- Fig. 6.25 upper part: FEFF and FDMNES disagree about the pre-edge position of CoO-hex. According to FEFF it is shifted about 1 eV to lower energies relative to that of CoO-cub, but according to FDMNES it is almost the opposite. Furthermore, according to FEFF CoO-hex shows a low-intensity whiteline at smaller energies than CoO-cub which along with the preferential pre-edge makes it a favored candidate for the site-specific shell.

- Fig. 6.25 lower part: For Co_3O_4 which contains Co^{3+} at the octahedral sites (like in CoO-cub) and Co^{2+} at the tetrahedral sites (like in CoO-hex) inside its cubic diamond structure, the contributions from both sites are simulated separately (see appendix C for details). They are shown

6.3. 1s3p-RIXS, HRFD-XANES and VTC-XES at undulator beamline ID26

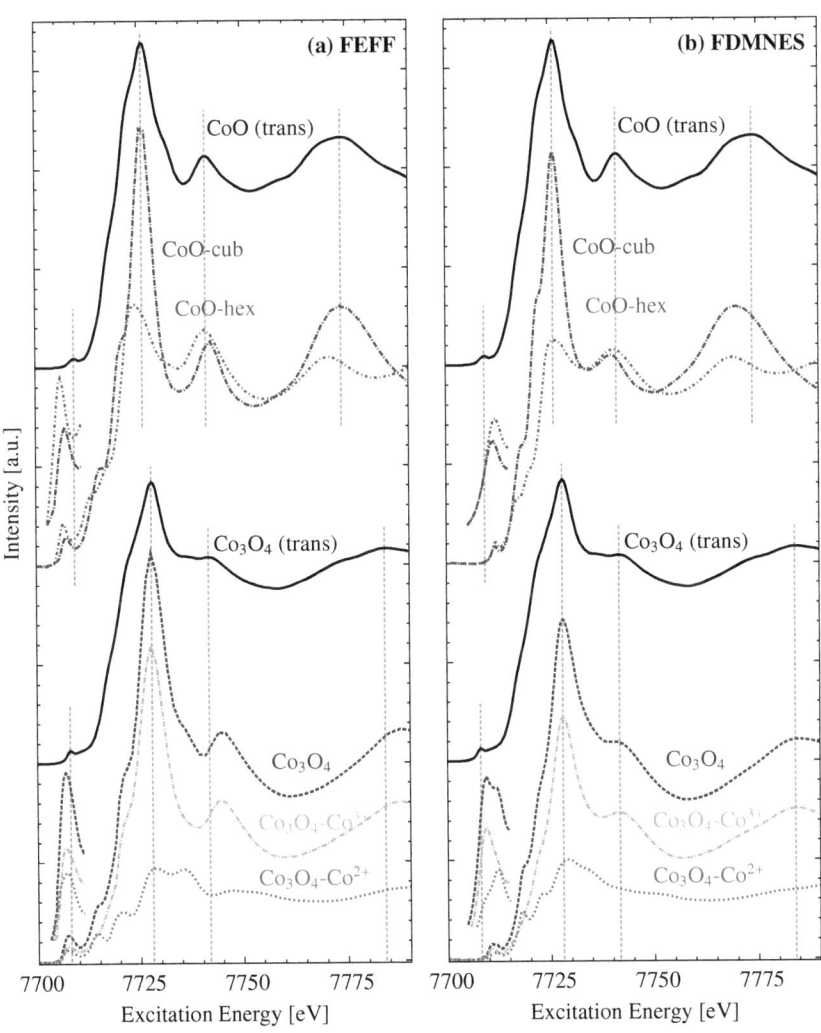

Figure 6.25.: Normalized Co K-edge XANES spectra of CoO and Co_3O_4 (both measured in transmission) along with respective simulations calculated by FEFF (a) and FDMNES (b). All pre-edge regions are also shown magnified whereby the position of the (center of) the peak is preserved, but its shape is stretched about a factor 2. Significant features of the experimental spectra are marked by dashed vertical lines.

6. Site-Selective XAS

with respect to their fraction in Co_3O_4 and additionally the sum of both is given. According to FDMNES the pre-edge of Co_3O_4 is composed of two energetically different contributions that belong to Co^{3+} (low energy) and Co^{2+} (high energy). This behavior is almost not visible in the FEFF simulations, it can just be surmised. Therefore no clear preference for the Co^{2+} or Co^{3+} part is possible. However, as it is known from the experimental Co_3O_4 spectrum that the pre-edge is (almost) suitable for the site-specific shell, but not the whiteline, the Co^{2+} part is preferable as its whiteline is less prominent.

All in all, a strong contribution by wurtzite CoO, as suggested earlier (see section 6.2.2), could not certainly be confirmed but is nonetheless, besides rocksalt CoO, the most likely possibility. An alternative explanation of the shell's low-energetic pre-edge is given by Co_3O_4 with mainly divalent Co. As the nanoparticles' shell allows only about 1 to 3 atomic layers, the occurrence of the spinel structure with a dominant Co^{2+} contribution is feasible. However, the whiteline of Co_3O_4, as well as that of its di- and trivalent components, is too high in energy, so that it can not be the dominant Co-oxide. It is obvious from the fit of the shell by CoO and $CoCO_3$ in Fig. 6.24 (a), however, that besides the Co-oxide other Co-compounds must be present in the nanoparticles shell. As one could only guess other Co-C compounds than $CoCO_3$ (or something different), this point remains open for now.

Site-selective XANES: more than two components

It is clear from the results until now that the nanoparticles' core spectrum can be well explained by the hcp phase, but that the shell is composed of at least two components: (1) CoO (rocksalt/wurtzite) and (2) $CoCO_3$. The LCF of the nanoparticles' $K\beta_{1,3}$ emission line showed no sensitivity to $CoCO_3$, which could be due to its resemblance to CoO with respect to NRXES, however, the quality factors of the XANES LCF in Table 6.6 showed a decrease by a factor two or more for the shell-rich Pos-3 of each nanoparticle, demanding for an additional fit component. Consequently, a LCF of the Co nanoparticles HRFD-XANES spectra (Fig. 6.19) by using Co-foil, CoO and $CoCO_3$ as fitting components is performed. The results are given in Table 6.8 and the R-factors for each fit are, in contrast to those of the 2-component fit, now all very similar.

Thus on can perform the SVD onto the nanoparticles' HRFD-XANES spectra and search for more than two theoretical site-selective spectra. This was not possible in the previous study (section 6.2.2) as a consequence of the lower resolution of the respective HRFD-XANES spectra, that led to "noise" for a third SVD component only. The Co : CoO : $CoCO_3$ ratios obtained by means of the LCF serve as an estimation for the ratios to be accommodated in the SVD. It is important here, however, that all sought site-selective spectra exhibit different valencies and thus a different chemical shift at the $K\beta_{1,3}$ emission line. Elsewise there would be no significant difference of their ratios at the three positions Pos-1, 2 and 3. In Fig. 6.17 it was recognizable that $CoCO_3$ has a slightly higher net valence spin than CoO, despite the coincidence of their formal oxidation state and from Table 6.8 it can be inferred that (a) the fractions of $CoCO_3$ are varying significantly from Pos-1 to 3 and (b) this variation is not strictly correlated to that of CoO. Consequently, is is tried to find at least two components of the nanoparticles' shell.

As in the previous SVD, all Pos-1 spectra of the nanoparticles will be excluded as they are improper due to their (by lifetime broadening) distorted pre-edge. A SVD of all remaining six spectra simultaneously will be performed as it has proven to yield the most reasonable results. It turned out again that it is impossible to get close to the ratios of the XANES LCF (see Table 6.8). Thus, instead it is tried to reproduce at least the trends of the ratios, i.e. core fraction decrease and shell fraction

6.3. 1s3p-RIXS, HRFD-XANES and VTC-XES at undulator beamline ID26

Table 6.8.: Ratios as obtained by LCF of the Co nanoparticles' HRFD-XANES spectra similar to Table 6.6, but with three fitting components Co, CoO and $CoCO_3$. The R-factors are describing the fit quality and the fit errors range between ± 0.02 ($R = 9$) and ± 1.0 ($R > 40$). SVD of the nanoparticles with spectra from Pos-2 and Pos-3 only: Listed are the ratios of the SVD with three (SVD-3) and with four (SVD-4) site-specific components core : shell-1 : shell-2 (: shell-3).

LCF XANES	R-factors [e-5] (Pos-1,2,3)	Pos-1	Pos-2	Pos-3
Co-nano-1	17, 9, 27	77.9 : 21.8 : 0.3	89.8 : 10.2 : 0.0	68.0 : 24.3 : 7.7
Co-nano-2	46, 46, 42	63.6 : 31.9 : 4.4	66.5 : 30.9 : 2.6	52.3 : 33.4 : 14.4
Co-nano-3	36, 39, 49	57.5 : 35.2 : 7.3	61.9 : 33.9 : 4.2	42.1 : 39.9 : 19.0
SVD-3	(Pos-2,3)		Pos-2	Pos-3
Co-nano-1	0, 9		93.4 : 5.3 : 1.3	54.1 : 24.7 : 21.2
Co-nano-2	2, 1		62.8 : 2.8 : 34.4	34.9 : 37.9 : 27.2
Co-nano-3	3, 2		58.8 : 11.0 : 30.2	31.6 : 64.6 : 3.8
SVD-4				
Co-nano-1	15, 2		100.0 : 0.0 : 0.0 : 0.0	45.0 : 15.1 : 19.1 : 20.9
Co-nano-2	3, 3		67.1 : 0.0 : 22.3 : 10.6	24.0 : 11.9 : 39.3 : 24.8
Co-nano-3	1, 4		62.2 : 0.0 : 27.5 : 10.3	18.6 : 19.3 : 44.4 : 17.7

increase, respectively, from Nano-1 Pos-2 to Nano-3 Pos-3. The calculated site-specific spectra are shown in Fig. 6.26 along with their model compounds and the ratios are given in Table 6.8.

- 3 site-specific components, Fig. 6.26 (a): With respect to the trends, the most suitable spectra have been obtained by SVD for the case that the core and shell-1 spectrum are (almost) identical to those of the 2-component SVD (compare Fig. 6.22). The third spectrum then became (not $CoCO_3$, but) almost structureless with a barely visible pre-edge at even lower energies (about 0.5 eV) than the shell-1 spectrum and an uncharacteristic whiteline at the same energy like shell-1. A closer look to the ratios in Table 6.8 reveals some questionable values though: The Pos-2 spectra of Co-nano-2 and 3 show a shell-2 fraction clearly larger than that of shell-1 which is conversely for the Pos-3 spectra. Lastly, Co-nano-3 at Pos-3 has almost no shell-2 contribution in contrast to the overall trend.

- 4 site-specific components, Fig. 6.26 (b): This time only the core spectrum had to be chosen identical to that of the 2-component SVD in order to get the best results. The resulting three shell spectra are different to all obtained before, in particular shell-1 exhibits features identical in energy to the common rocksalt CoO. Shell-2 has a unusual broad pre-edge that centers at about 7710 eV and a whiteline position at 7724.8 similar to $CoCO_3$. The shell-3 spectrum eventually shows the low-energetic pre-edge which was present in all previous shell spectra, however, otherwise it is different to these. As it was very noisy, a smoothed spectrum of it is also given. To control the reasonability of these site-specific spectra the ratios will be consulted (Table 6.8): At first, the R-factor for Co-nano-1 at Pos-2 seems unexpectedly high and of course this spectrum is for sure not 100% core, anyway, as can be seen in the SVD-3, Co-nano-1 at Pos-3 also constitutes a high R-factor, so actually there is always one outlier. Unexpected, but

6. Site-Selective XAS

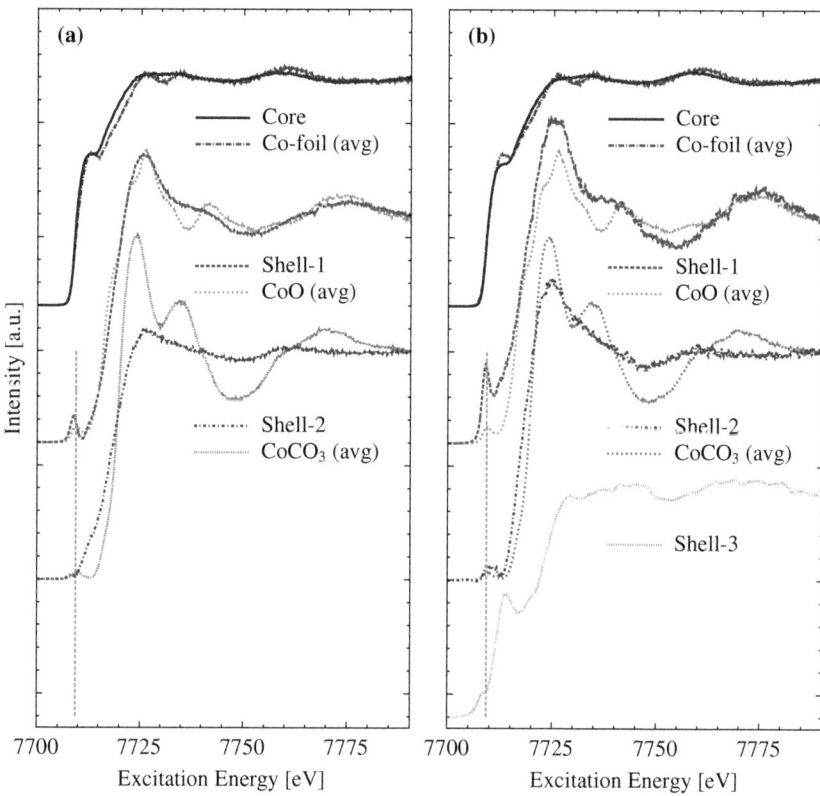

Figure 6.26.: Site-specific Co K-edge XANES spectra of the Co nanoparticles as obtained by SVD with 3 components (a) and with 4 components (b). The Co references are also plotted. The vertical dashed line marks the pre-edge position of CoO (avg) at 7709.3 eV.

nonetheless possible, is that all Pos-2 spectra show no shell-1 contribution and that the core fractions of Co-nano-2 and 3 at Pos-3 are the very low.

All in all, the extension of the SVD to resemble the real number (not of sites, but) of Co compounds with different valencies (or more precise: net valence spin) inherent in the Co nanoparticles, has proven to be promisingly. Especially the results of the 4-component SVD yielded feasible ratio trends and reasonable spectra. However, albeit it will be shown (in section 6.3.6) that besides oxygen and carbon also nitrogen is ligated to Co inside the nanoparticles, there is no substantial information about the particular crystallographic structure of these compounds. Thus, one would have to guess, out of a vast number, which simulations to perform to identify the site-specific XANES spectra gained (and shown in the right panel of Fig. 6.26). As these simulations are moreover rarely identical but only similar to the real experimental spectra, it would be mandatory at his point, in order to proceed, to measure suitable references.

6.3.5. Valency/Site-selective EXAFS

Now it comes to the investigation of the HRFD-EXAFS spectra. The references, recorded at their respective $K\beta_{1,3}$ peak positions, are shown in Fig. 6.27 and the nanoparticles in Fig. 6.28. The latter are recorded at $K\beta_{1,3}$ fluorescence positions Pos-1, 2 and 3 identical to those in the previous section (see Table 6.6). For all samples the k^2-weighted k-space spectra are given on the left and the magnitudes of the R-space spectra, which are fourier transformations of the k-space spectra, on the right. The fourier transformation is performed within the k-range 3 to 11 $Å^{-1}$, i.e. 7743.3 to 8170.0) eV. Hereby, the lower limit corresponds to a position a few eV above the whiteline of metallic Co and the first shape resonance of CoO and $CoCO_3$, respectively, to assure that no bound states are involved (since the EXAFS investigation relies on the analysis of continuum states). The upper limit gives the borderline where the data quality gets too worse, in particular for the Pos-3 spectra.

According to the EXAFS equation Eq. (2.33), the peaks in R-space are the result of the surrounding atoms that, especially in a crystalline material, are positioned at (more or less) discrete distances (given rise to coordination-shells), which are given relative to the absorbing atom, positioned at the origin, i.e. at 0 Å. However, the positions of the peaks are not the real distances as a R-space spectrum is only the sum of all scattering paths to these coordination-shells. These single paths are oscillating functions in R-space with a real and imaginary part and interfere constructively or destructively with each other. Furthermore, they are phase-shifted by the potential of the absorbing atom. Thus, the real distances are obtainable only via appropriate simulations of the various electron scattering paths fitted to the experimental spectrum (see appendix D for more details).

From the R-space spectra of the references in Fig. 6.27 it is evident that the first coordination-shell at about 1.7 Å is specific for CoO and $CoCO_3$ and thus can be attributed to light elements (C and O) and the second one at about 2.2 Å to Co atoms as it is the first for Co-foil. Consequently, for the nanoparticles in Fig. 6.28 the partial site-selectivity is showing up in the increase and decrease, respectively, of the first and second coordination-shell peak when going from Pos-2 to 1 to 3. Here the most drastic changes are visible for the Pos-3 spectra in the whole k-range shown and between 1 and 3 Å in R-space (thereafter the oscillations are similar for all positions, the peaks are just weaker and less sharp).

All nine nanoparticles k-space spectra with k-weight = 0 are fitted by the references Co, CoO and $CoCO_3$ that were shown in Fig. 6.27. The resulting ratios are given in Table 6.9 (the fitting curves are not shown). The fit quality for k-space EXAFS is worse about three orders of magnitude which should be a strong argument to discard these fits. However, the fits are only poor with respect to the amplitudes, the oscillations are all reproduced. The intensity decrease of the amplitudes stems from the fact that due to the small size of the nanoparticles, the fraction of surface atoms is significant in contrast to the bulk references that are used for the fit. These surface atoms exhibit a reduced number of neightbors though and hence the net coordination numbers of the nanoparticles, to which the EXAFS oscillation amplitudes are proportional, are somewhat reduced. This effect showed already up in the shape resonances of the HRFD-XANES spectra (see Fig. 6.22).

However, contrary to the LCF's of the nanoparticles' $K\beta_{1,3}$ and the respective HRFD-XANES spectra (see Table 6.6 and 6.8), Co-nano-2 now shows the highest Co metal contribution – which could already be recognized in Fig. 6.28 as Co-nano-2 exhibits the strongest amplitudes in k- and R-space. To be precise, for EXAFS the amplitudes of the three Pos-2 spectra were expected to be almost identical, as the fraction of surface atoms of the respective metal cores are almost identical with respect to the estimated core diameters for Nano-1, 2 and 3 (46, 51 and 53 Å, see section 6.3.3). Obviously, the EXAFS spectra are falsified by some measuring problems, most probably by self-absorption (SA) effects due to too thick samples, which make the spectra useless with respect to

6. Site-Selective XAS

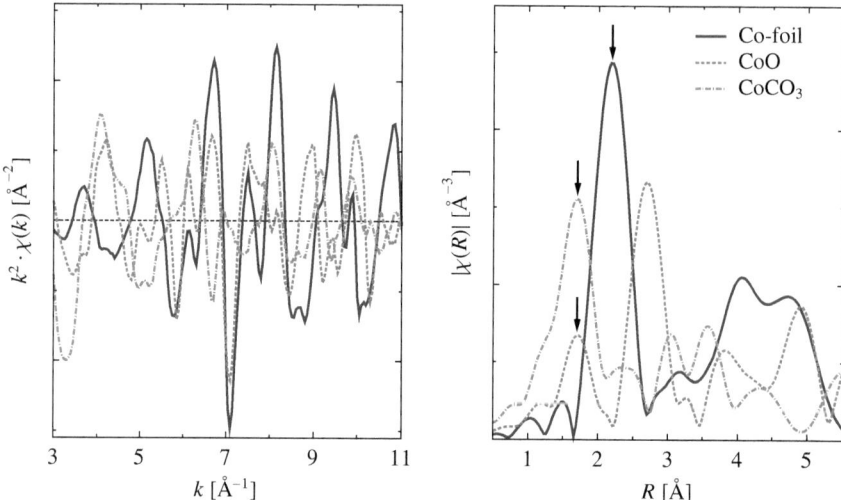

Figure 6.27.: Left: k^2-weighted k-space EXAFS spectra of the references Co-foil, CoO and CoCO$_3$. Right: Phase-corrected magnitudes of the R-space EXAFS spectra, fourier transformed from k-space spectra on the left. The labels are denoting the spectra on the left likewise.

the amplitudes and thus, to the coordination numbers (in the interpretation). Although sophisticated algorithms for the correction of self-absorption effects in EXAFS do exist, e.g. by C. H. Booth and F. Bridges [13], they still do not allow for unambiguous corrections in general. It will be searched for the pure site-selective spectra nonetheless, as there can be drawn valuable information from the EXAFS oscillations, i.e. distances, lattice constants and to a certain degree even thermal and static disorder, owing to the Debye-Waller factors.

The SVD is performed via method 1, i.e. a SVD for each nanoparticle separately, as well as simultaneously for all nine HRFD-EXAFS spectra. The results from the simultaneous method ignore the fact, however, that the local structure is expected to be different at least for the shell of the three nanoparticles, which varies between 1 and 3 monolayers. Method 2 is discarded, since it gives unreasonable results again: the site-selective core spectrum appears correct in the XANES region but exhibits EXAFS oscillations of a divalent material and for the site-selective shell spectrum it is vice versa. The number of site-specific spectra is restricted to two for the SVD, i.e. a metallic core and one shell, since for EXAFS it is generally not possible to distinguish between the different first period ligands C, O or others that are expected to be present inside the nanoparticles. Furthermore, as can be seen in the left panel of Fig. 6.28, the quality of the nanoparticles' Pos-3 spectra is rather poor so that one can not expect to get more than one valuable spectrum. Consequently, the to be determined shell spectrum will comprise all shell components, i.e. it will reflect contributions from Co-O and Co-C (and other) compounds at the same time.

The resulting ratios of the SVD's are given in Table 6.9. For method 1 it was possible to adjust the core's SVD ratios to the Co-foil ratios from the k-space EXAFS LCF within the given errors. For the method S by contrast the limits of the core fraction with respect to Co-nano-1 at Pos-2 as well as the limit of the shell spectrum with respect to the energy axis had to be chosen, in order to get ratios that

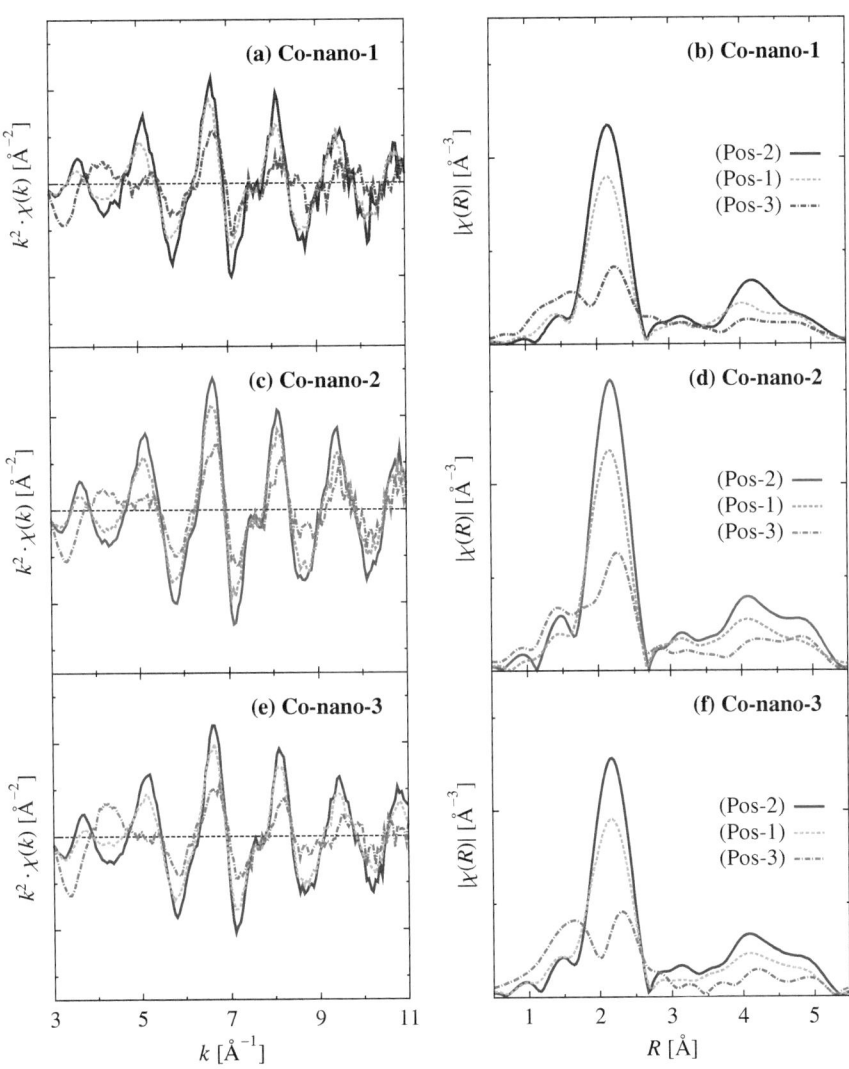

Figure 6.28.: Left: k^2-weighted k-space HRFD-EXAFS spectra of the nanoparticles Co-nano-1, 2 and 3. Right: Magnitudes of the R-space HRFD-EXAFS spectra, fourier transformed from k-space spectra on the left. The labels are denoting the extraction position for the spectra (on the left and right likewise).

6. Site-Selective XAS

Table 6.9.: LCF of the Co nanoparticles' HRFD-EXAFS k-space spectra (k-weight = 0) in the k range 3 – 12 Å^{-1} with three fitting components Co, CoO and CoCO$_3$. The ratio errors are between ±2 and ±4, depending on the R-factors which are given for each three fits in a column. Below are the core : shell ratios as determined by two different SVD's.

LCF EXAFS	Overall	Pos-1	Pos-2	Pos-3
Co-nano-1	75.1 : 20.0 : 4.9	71.1 : 27.7 : 1.2	79.3 : 20.3 : 0.4	47.3 : 37.5 : 15.2
Co-nano-2	77.2 : 14.8 : 8.0	74.7 : 21.8 : 3.5	84.8 : 12.0 : 3.3	48.9 : 33.5 : 17.5
Co-nano-3	68.9 : 21.4 : 9.7	67.2 : 26.1 : 6.7	78.6 : 17.6 : 3.8	39.6 : 40.8 : 19.7
R-factors [e-3]	1554, 578, 1061	2424, 996, 1297	944, 348, 641	1186, 667, 597
SVD (method 1)				
Co-nano-1	–	71.8 : 28.2	77.8 : 22.2	46.6 : 53.4
Co-nano-2	–	75.3 : 24.7	83.3 : 16.7	48.7 : 51.3
Co-nano-3	–	66.6 : 33.4	76.6 : 23.4	39.5 : 60.5
R-factors [e-3]	–	65, 45, 24	26, 13, 10	19, 13, 7
SVD (method S))				
Co-nano-1	–	78.4 : 21.6	87.6 : 12.4	38.2 : 61.8
Co-nano-2	–	79.1 : 20.9	89.4 : 10.6	37.0 : 63.0
Co-nano-3	–	67.9 : 32.1	82.7 : 17.3	25.1 : 74.9
R-factors [e-3]	–	136, 28, 27	26, 67, 27	64, 14, 9

as best as possible fit to those of the k-space EXAFS LCF. Here it is important to mention that for method 1 (in contrast to the method S) further variation of the SVD parameters would be possible. Thus, for the case of, e.g., shell reference spectra different to CoO and CoCO$_3$ and consequently different EXAFS LCF ratios, deviating site-specific components would be feasible. The site-specific EXAFS spectra resulting from the SVD's are shown in Fig. 6.29. All spectra are k^2-weighted and the fourier transformation is performed in the k-range from 3 to 11 reciprocal Å. The core spectra from both methods show coincident coordination-shells, just the amplitude of the first-shell peak at about 2.2 Å (not the real distance) is varying as are the amplitudes in k-space and especially for the core from method S it is significantly smaller. The shell spectra by contrast are more different with respect to the two methods visible in k and R-space. To investigate this in more detail, the the magnitude and the real part of the R-space spectra of the simultaneously determined shell, the averaged shell of method 1 and the average of the three nanoparticles from Pos-3 are compared in Fig. 6.30 (c+d). Obviously, the averaged shell spectrum from method 1 is completely different to the "shell-rich" Pos-3 spectrum and to the shell spectrum of method S, which seems dubious and will be clarified in the fit. The respective spectra are also shown for the core and Pos-2 of the nanoparticles in Fig. 6.30 (a+b) to demonstrate that for the core both methods are (almost) coincident. Furthermore, it can be seen that the core spectrum from method S is almost identical to the average of the nanoparticles' spectra from Pos-2 (within about 10 %, see Table 6.3). As a consequence, the Pos-2 and Pos-3 spectra will also be fitted to get information on the core and shell, respectively.

EXAFS fits: Core

In order to find the correct Co-phase for the site-specific core spectrum, the four known phases (see Table C.3 for crystallographic details) are simulated by FEFF and compared to the core determined

6.3. 1s3p-RIXS, HRFD-XANES and VTC-XES at undulator beamline ID26

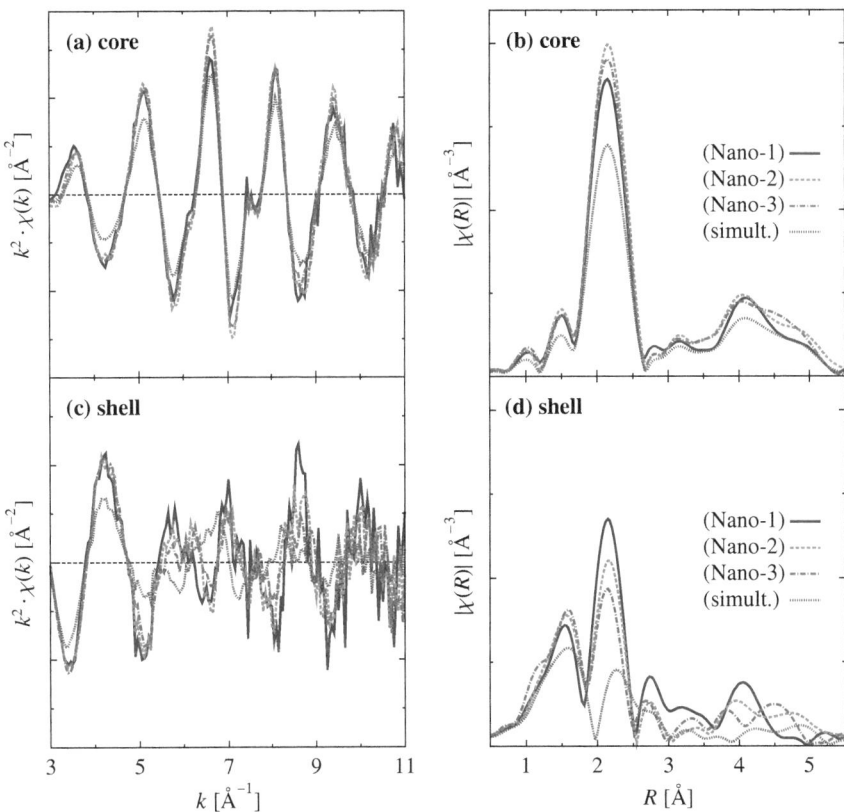

Figure 6.29.: Site-specific EXAFS spectra for the core and shell as determined by the SVD. The labels are denoting the methods, as explained in the text, for the left and right spectra likewise. The k^2-weighted k-space spectra on the left are fourier transformed to give the (magnitudes of the) R-space spectra on the right.

by method S in Fig. 6.31. The simulations are, according to the EXAFS equation Eq. (2.33) or Eq. (D.2), the sum of all paths, up to triple-scattering, within a cluster of radius 8 Å (see appendix D for more details). With respect to both k and R-space the hcp crystal structure is the most suitable for certain, though fcc is possible too, since it differs from hcp first at the 3rd and 4th coordination-shell (at about 3.9 Å and 4.8 Å), which occurs from an enhanced number of surrounding atoms at these distances as a consequence of the fcc lattice. Thus, both phases will be simulated and fitted path by path to the various core spectra and the nanoparticles' Pos-2 HRFD-EXAFS spectra. A simultaneous fit by both phases is not possible, however, as most of the the hcp paths are identical to those of fcc, and thus the high correlation would inhibit a reliable result. The fits are performed in R-space from 1 to 5 Å with k-weights of 1, 2 and 3 simultaneously (to regard the whole k-range likewise) and are shown for the core of method S in Fig. 6.32.

6. Site-Selective XAS

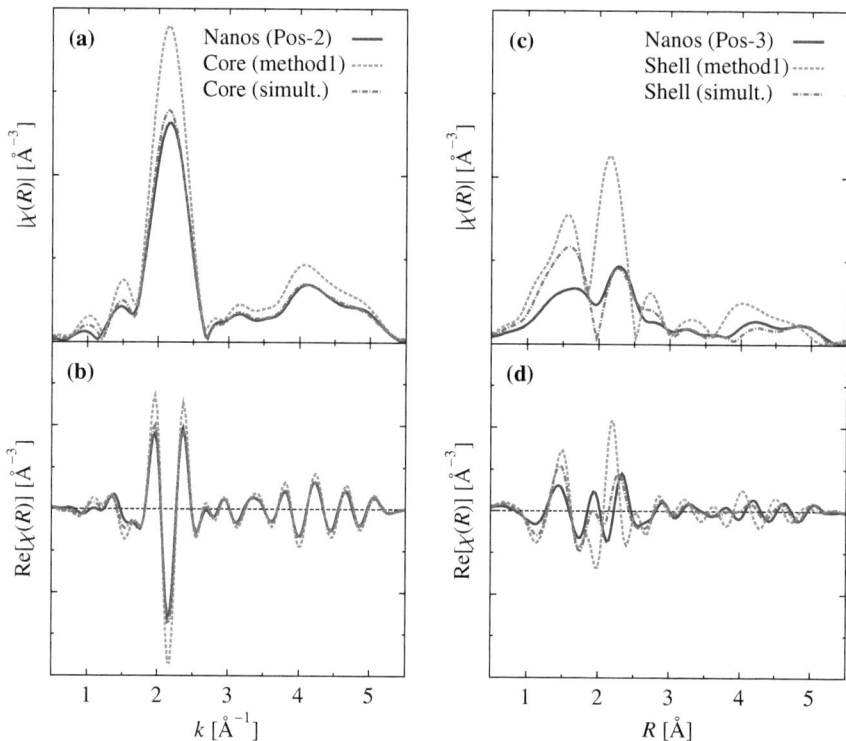

Figure 6.30.: Left (Right): Site-specific EXAFS spectra for the core (shell) in comparison to the nanoparticles' HRFD-EXAFS spectra averaged at Pos-2 (Pos-3). The core (shell) spectrum of method 1 is also averaged. Shown is the magnitude of $\chi(R)$ on top and its real part at the bottom.

Besides the single-scattering (ss) paths shown in this figure, all double- (ds) and triple-scattering (ts) paths with significant amplitudes have also been included in the fit. In particular for the 4th coordination-shells there are strong collinear ds and ts paths likewise for hcp and fcc. The detailed results are given in Table 6.10 along with those from EXAFS fits of Co-foil, the three nanoparticles from Pos-2 (fitted simultaneously) and also the three cores from method 1 (fitted simultaneously). The latter fit is consistent with the other results mainly, but has to be treated with care nonetheless, as the corresponding shell spectra are unreasonable (*vide infra*). The distances R_n ($n = 1 - 4$) were not allowed to run freely in the fit, but were restricted to their respective distances defined by the crystallographic lattice (also given in Table 6.10). Instead the lattice constants (a and c for hcp, a for fcc) were taken as parameters. Thereby, it was ensured that really the suitability of the hcp or fcc phase was tested and the lattice constants (and coordination-shell distances likewise) could be determined accurately. Further parameters have been a constant energy shift E_0, a reduction factor δN_n and four Debye-Waller factors σ^2, one for each main coordination-shell. δN reflects the total amplitude

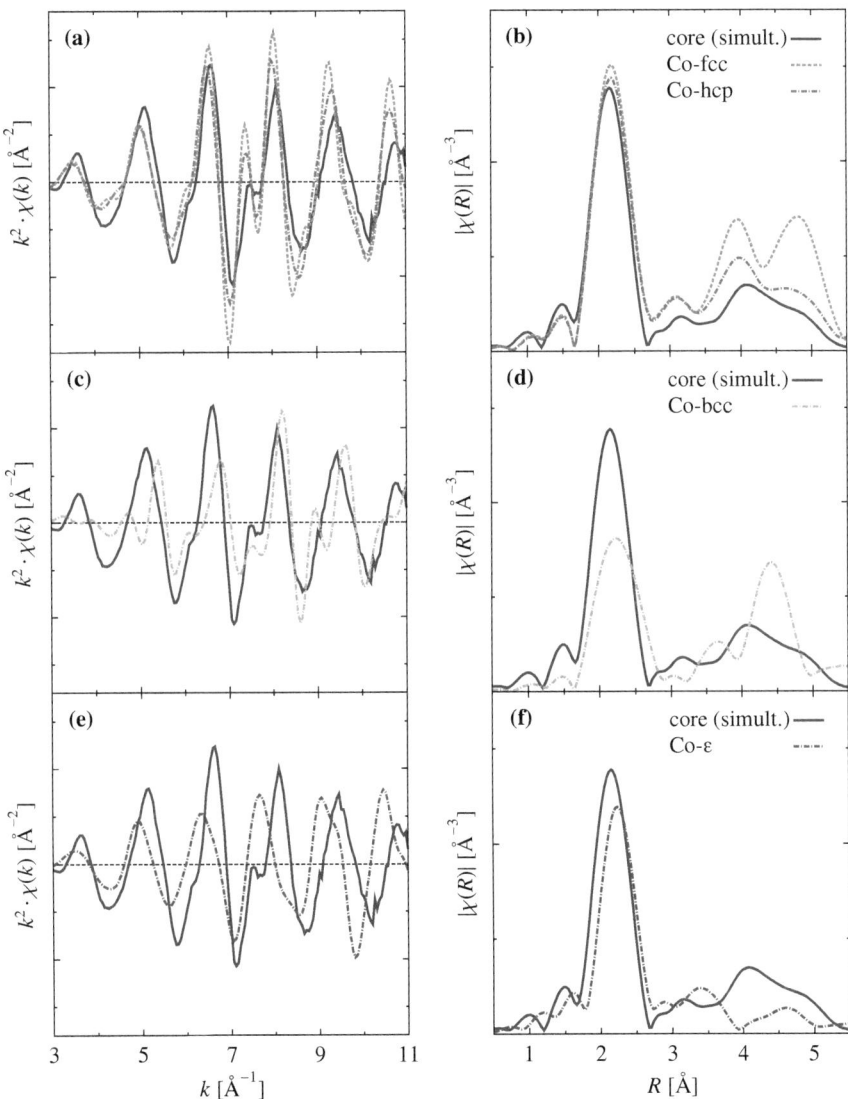

Figure 6.31.: Site-specific EXAFS spectrum of the core of method S in comparison to the spectra of the FEFF simulated metal Co-phases (no fit, but downscaled by factor 0.15) in k-space (left) with k^2-weight and in R-space (right) fourier transformed from spectra on the left.

6. Site-Selective XAS

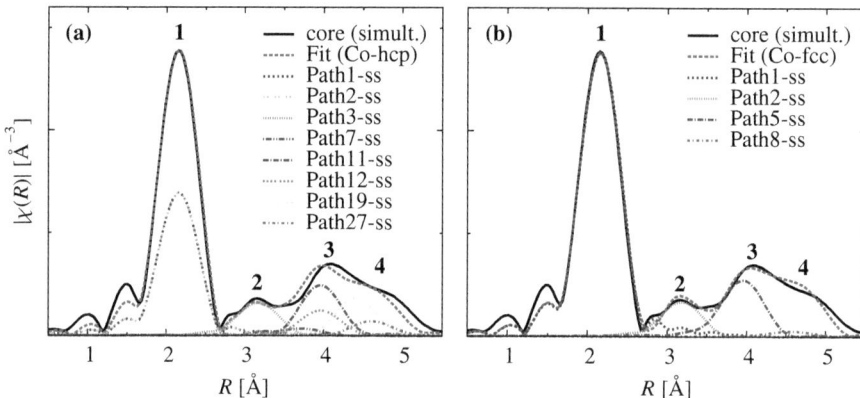

Figure 6.32.: Core spectrum from method S fitted by FeIT simulation of Co-hcp (left) and Co-fcc (right). Also shown are the single-scattering (ss) paths, which reflect the coordination-shells, marked by boldface numbers.

reduction of the coordination of each shell due to (1) the high surface fraction of the nanoparticles and (2) as a result of the measurement itself (e.g. it accounts for the mentioned SA effects, etc). Those two effects can not be disentangled and of course they can also modify the σ^2's – which are somehow correlated with δN – but to a small degree only, since the σ^2's not only determine the amplitude, but also the width of the oscillations (remember the factor $exp(-2k^2\sigma^2)$ in the EXAFS equation Eq. 2.33 or Eq. D.2). In total there are 8 (7) parameters for the hcp (fcc) fit, which are fitted to the spectra that exhibit 20 independent points, in accord with Eq. (D.3) (see appendix D).

In order to have a reference and to recognize trends starting from a perfect crystal to the nanoparticles and its core, Co-foil is also fitted by Co-hcp and Co-fcc. In opposition to XANES (see appendix C), Co-foil reveals itself to be purely Co-fcc with respect to the local (EXAFS) structure. Its hcp fit consequently yields unphysical values for the σ^2's of the 3rd and 4th coordination-shell, as the hcp amplitudes are about 50 % and 100 % weaker compared to those of fcc. Nonetheless, this fit is useful regarding the oscillations and consequently lattice constants and distances.

As can be recognized visually (Fig. 6.32) and via the R-factors (Table 6.10), the hcp-phase is a proper choice for the Co nanoparticles' core(s) and also for their Pos-2 spectra. The R-factor is worse, however, for Nano-123 and for Core-123 about a factor 2 compared to the core from method S, which is understandable for the former as the Co-nano spectra are still inhibiting a (small) contribution from the divalent shell, for the latter though, it is not reasonable. From the Debye-Waller factors it can be inferred moreover that the metallic core of the nanoparticles is exclusively hcp (with respect to the local structure), since (in contrast to Co-foil) the σ^2 are increasing from coordination-shell 1 to 3 to 4 (coordination-shell 2 is an exception, see also fits of Co-foil and Co-powder measured at ANKA in appendix D and Table D.2), as it should be the case for a mono-atomic material. A decrease of σ_4^2 in the current case would indicate a significant contribution from fcc as its 4th coordination-shell amplitude is about two times that of hcp (as it is the case for Co-foil). It is important to note here that the nanoparticles' core can be stated highly crystalline, as in contrast to the EXAFS fits of the Co-Pt catalysts in section 5.4, a rigid hcp lattice (remember that only the lattice constants were allowed to vary) was successfully fitted.

6.3. 1s3p-RIXS, HRFD-XANES and VTC-XES at undulator beamline ID26

Table 6.10.: EXAFS fits of HRFD-EXAFS spectra of Co-foil, of Co-nano-1+2+3 from Pos-2 simultaneously, of the site-specific core-1+2+3 from method 1 simultaneously and of the site-specific core from method S by FEFF simulations of Co-hcp (top) and Co-fcc (bottom).

	perfect crystal	Co-foil (ESRF)	Nano-123 (Pos-2)	Core-123 method 1	Core method S
HCP ::: R-factor [$\times 10^{-3}$]	–	24	14	16	8
δN	1.00	1.0(1)	0.66(4)	1.0(1)	0.73(5)
a [Å]	2.507	2.503(09)	2.495(04)	2.490(04)	2.489(06)
c [Å]	4.070	4.075(27)	4.070(13)	4.065(14)	4.065(18)
$R_1 = \sqrt{a^2/3 + c^2/2}$ [Å]	2.502	2.501(12)	2.494(6)	2.489(6)	2.489(08)
$R_2 = \sqrt{4a^2/3 + c^2/2}$ [Å]	3.538	3.537(16)	3.527(8)	3.521(8)	3.520(11)
$R_3 = \sqrt{7a^2/3 + c^2/2}$ [Å]	4.340	4.334(15)	4.321(7)	4.312(7)	4.311(10)
$R_4 = 2a$ [Å]	5.014	5.007(17)	4.990(8)	4.979(8)	4.978(11)
σ_1^2 [10^{-3}Å^2]	–	3(1)	5.2(7)	5.8(5)	5.6(6)
σ_2^2 [10^{-3}Å^2]	–	5(3)	11(3)	11(2)	11(2)
σ_3^2 [10^{-3}Å^2]	–	1(1)	7(1)	8(1)	8(1)
σ_4^2 [10^{-3}Å^2]	–	0(2)	8(4)	10(3)	9(4)
FCC ::: R-factor [$\times 10^{-3}$]	–	7	18	20	11
δN	1.00	0.91(6)	0.64(4)	0.99(6)	0.71(5)
a [Å]	3.544	3.536(5)	3.528(4)	3.522(5)	3.521(6)
$R_1 = a\sqrt{1/2}$ [Å]	2.506	2.501(3)	2.495(3)	2.490(3)	2.490(4)
$R_2 = a$ [Å]	3.544	3.536(5)	3.528(4)	3.522(5)	3.521(6)
$R_3 = a\sqrt{3/2}$ [Å]	4.341	4.331(6)	4.321(5)	4.313(6)	4.312(8)
$R_4 = a\sqrt{2}$ [Å]	5.012	5.001(7)	4.989(6)	4.980(6)	4.979(9)
σ_1^2 [10^{-3}Å^2]	–	2.6(05)	4.8(6)	5.6(5)	5.4(7)
σ_2^2 [10^{-3}Å^2]	–	3.6(13)	10(3)	11(2)	10(3)
σ_3^2 [10^{-3}Å^2]	–	3.0(07)	12(2)	13(2)	13(2)
σ_4^2 [10^{-3}Å^2]	–	3.5(19)	25(5)	29(5)	27(5)

The Co-fcc fits seem to be of similar quality regarding the R-factors. However, a closer look to the σ^2's reveals unphysical huge values for the 4th coordination-shell as the respective amplitude of fcc is much too intense, which once more confirms the (dominance of the) hcp-phase. The fcc fits are nevertheless useful, since the obtained shell distances R_n are confirming (with higher accuracy even) those of the hcp fit. Furthermore, the presence of Co-fcc in the nanoparticles' core is still possible, due to the congruence of both phases to a certain degree (see Fig. 6.31).

Eventually a trend is visible regarding the lattice constants a and c (and also a of a hypothetical fcc phase contribution) and the respective coordination-shell distances R_n: They are decreasing slightly from a perfect crystal and the Co-foil, respectively, to the Co nanoparticles Pos-2 spectra and to the site-specific core spectra. However, this effect is vanishingly low, smaller than 1 %, thus despite the small size of the nanoparticles there is no significant lattice contraction.

6. Site-Selective XAS

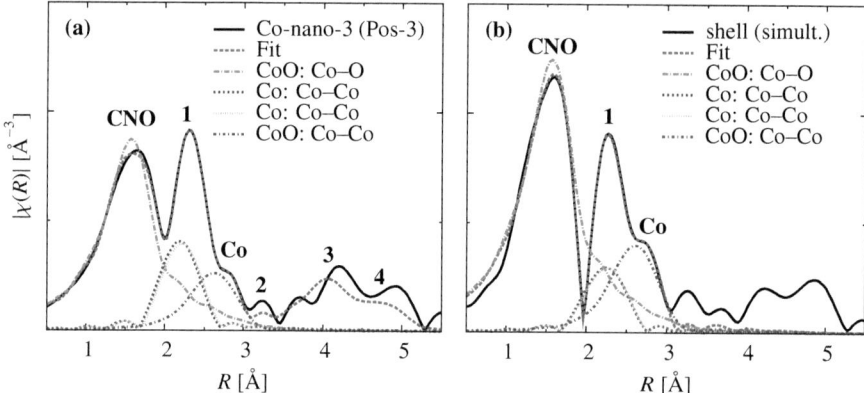

Figure 6.33.: HRFD-EXAFS spectrum (left) and site-specific shell spectrum (right) from method S fitted by Feff simulation of rocksalt CoO and Co-hcp. Also shown are the single-scattering (ss) paths, which reflect the coordination-shells, marked by boldface labels/numbers.

Lastly, it has to be mentioned that additionally all nine experimental HRFD-EXAFS spectra of the Co nanoparticles (three types and three positions) have been fitted separately, in order to find correlations of core properties with the shell thickness. However, no significant trends could be recognized, which indicates that the nanoparticles' shell has no discernible influence onto the properties of the complete nanoparticle, with respect to the viewpoint of EXAFS and the range of available shell thicknesses.

EXAFS fits: Shell

As the divalent shell of the nanoparticles was estimated to be very thin, between 3.5 and 7.0 Å (see section 6.3.3), only cluster radii of 3.5 Å at the maximum are conceivable and significant EXAFS structures should be visible only up to $\simeq 3.5$ Å also. The adjoining zerovalent metal core is contributing in the whole range by contrast, even its first coordination-shell, owing to the Co atoms placed at the inner surface of the nanoparticles' shell. Above $\simeq 3$ Å pure Co metal is expected neither, as the shell is not symmetric with respect to all three dimensions (as it is a sphere's layer) and what leads to the following: When going along so-called "small circles" of the nanoparticle sphere – instead of along the radial direction – it is possible to have clusters with average radii of 10 Å (half chord of a circle), however, only in two dimensions as the radial radius of such a cluster is still restricted by half of the shell's thickness (1.8 to 3.5 Å). Moreover, only the clusters centered inside the shell layer are that huge and in total, contributions from the shell above 3.5 Å are expected to be weak and diffuse.

Thus, the site-specific shell from method S as well as the three shells from method 1 (simultaneous fit) and the HRFD-EXAFS spectra of Co-nano-1, 2 and 3 from Pos-3 (simultaneous fit) are fitted by a FEFF simulation of Co-hcp with an identical parameter set as in the previous section to account for the metal part and additionally by a FEFF simulation of a Co–O and a Co–Co path (like in any Co-O compound), whereby both paths distances are free parameters. The two new paths also have one common amplitude reduction factor δN_O and each a Debye-Waller factor σ^2. The fits are shown

6.3. 1s3p-RIXS, HRFD-XANES and VTC-XES at undulator beamline ID26

Table 6.11.: EXAFS fits of the HRFD-EXAFS spectra of Co-nano-1, 2, 3 from Pos-3 simultaneously and of the site-specific shell from method S by FEFF simulations of Co-hcp and Co-O.

		perfect crystal	Nano-123 (Pos-3)	Shell method S
	R-factor [$\times 10^{-3}$]	–	33	4
	δN	1.00	0.22(4)	0.12(6)
Co metal (hcp)	R_1 [Å]	2.50	2.49(1)	2.48(2)
	R_2 [Å]	3.54	3.52(1)	–
	R_3 [Å]	4.34	4.33(2)	–
	R_4 [Å]	5.01	5.01(2)	–
	σ_1^2 [10^{-3}Å^2]	–	5(2)	3(4)
	σ_2^2 [10^{-3}Å^2]	–	10(5)	–
	σ_3^2 [10^{-3}Å^2]	–	6(2)	–
	σ_4^2 [10^{-3}Å^2]	–	2(12)	–
Co-O	δN_O	1.00	0.63(7)	1.1(1)
	R_{CNO} [Å]	–	2.00(1)	2.00(1)
	R_{Co} [Å]	–	2.99(2)	2.99(2)
	σ_{CNO}^2 [10^{-3}Å^2]	–	9(2)	10(3)
	σ_{Co}^2 [10^{-3}Å^2]	–	20(5)	18(5)

in Fig. 6.33 and the detailed results in Table 6.11, though not for the method 1 shell spectra that turned out to be impossible to fit, even not when increasing the number of parameters, e.g. allowing all shell distances to float. The reason for the failure of method 1 is most probably the insufficient quality of a set of three HRFD-EXFAS spectra in contrast to nine for method S. The shell spectrum of method S could be well fitted up to 3 Å, but afterwards the fit quality became unacceptable, so that the respective fit was restricted to 3 Å. Actually the site-specific shell spectrum might not contain any Co metal as it should have been separated out (and attributed to the site-specific core) in the SVD process. However, obviously at least the 1st coordination-shell of Co-hcp (labelled "1") is clearly present.

For Co-nano-123 at Pos-3, it was possible to fit all the hcp coordination-shells 1 to 4, but with significant disturbances by the diffuse shell contributions above 3 Å, as can be seen for Co-nano-3 in Fig. 6.33 (a). An attempt to simulate these diffuse coordination-shells failed, however. Anyway, the initially presumed simple core-shell model is confirmed by these results. Further on, it was searched for trends in the three Pos-3 spectra with respect to lattice constants, bond distances and Debye-Waller factors. The aim of this was to find relations between these local structure properties and the shell-thickness, which is increasing from Nano-1 to 3. However, all properties were, within the fitting errors, almost coincident.

The coordination-shells of the site-specific shell, labelled "CNO", "1" and "Co", can be clearly recognized and yield distances R_{CNO} = 2.00(1) Å, R_1 = 2.48(2) Å and R_{CNO} = 2.99(2) Å. Here CNO indicates that all three first period elements C, N and O could be present in the nanoparticles' shell at this distance, as they are indistinguishable in EXAFS and "Co" stands for a further coordination-shell of Co atoms. In comparison to previous fits the disorder of the two CoO coordination-shells

6. Site-Selective XAS

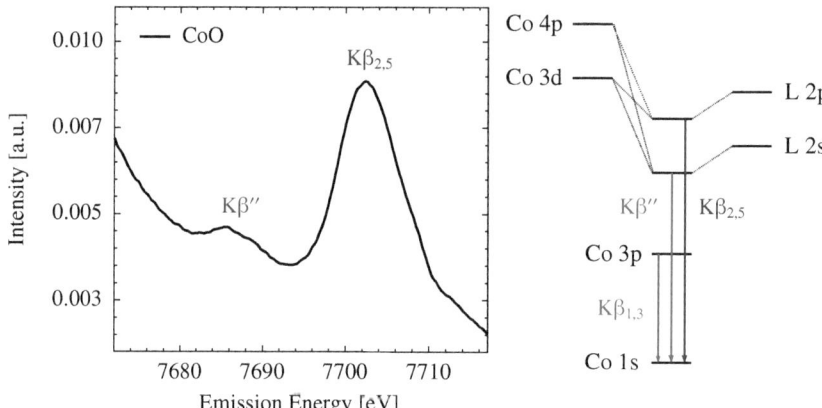

Figure 6.34.: VTC-XES spectrum of CoO with labels for the peaks on the left and which origins are schematized at the right ("L" denotes the ligand).

(CNO and Co), given by σ^2, is strong, especially for the CoO-like Co-shell, which is understandable as only a low degree of crystallinity is possible in such a thin shell due to the contact with different "environments" on both surfaces, inner and outer. The three coordination-shells, identified in the site-specific shell spectrum, are only separated by about 0.5 Å in each case. Consequently, at the minimum there have to be two distinct phases present in the nanoparticles' shell, as no two Co coordination-shells can coexist that close in one phase, or to be precise, one shell-phase and the remnant of the core's hcp-phase. Thus, the two phases Co-hcp and CoO used for the FEFF simulations can assumed to be present in the site-specific shell spectrum.

However, the first two coordination-shells of a rocksalt CoO are located at 2.133 Å and 3.017 Å (see Table C.3 for crystallographic details) and hence only the second is suitable for the site-specific shell. Wurtzite CoO on the other hand has a proper 1st coordination-shell distance of 1.980 Å, but improper second one at 3.206 Å, according to FEFF simulations. Consequently, as there are strong indications from the other measurements (see next section 6.3.6) for the presence of C and N, whose paths would overlap with that of O, the determined distance of $R_{CNO} = 2.00(1)$ Å can be explained as the average distance of the three contributions.

6.3.6. Valence-to-core spectroscopy (K$\beta_{2,5}$)

In this section, additional light will be shed onto the kind of the ligands contributing to the shell component of the Co nanoparticles. Identifying the metal ligands usually has to be performed at quite low energies (O K-edge and L-edge at 543.1 eV and 41.6 eV, respectively), e.g., via ultraviolet and visible (UV-vis) or via infra-red spectroscopy (IR), which all give characteristic information on the ligands. However, for UV-vis spectroscopy high vacuum conditions are required, which poses some limitations to the samples under investigation, and IR spectroscopy is only capable of detecting covalently bound ligands, which is not known a priori in most cases.

It is well known for several years that X-ray emission below the metal Fermi level, after creation

6.3. 1s3p-RIXS, HRFD-XANES and VTC-XES at undulator beamline ID26

of a 1s core hole, yields the $K\beta''$ and $K\beta_{2,5}$ features (see left panel of Fig. 6.34 for CoO) that reflect the valence-to-core (VTC) transitions and thus also gives information about the ligands. Indeed, for transition metals in general, the assignment of the $K\beta''$ "crossover-peak" with metal 2s to ligand 1s transitions was well known for quite some time (see Ref. [94]). Since the VTC-XES (or $K\beta_{2,5}$-XES) transitions are in the hard X-ray range, no special vacuum conditions are necessary. However, as was demonstrated in Fig. 6.15 such measurements suffer from low intensity about two orders of magnitude compared to the main $K\beta_{1,3}$ emission line. With the onset of 3rd generation synchrotron sources, providing higher photon flux from insertion devices and due to the development of high-resolution spectrometers, the lack of intensity is no longer a problem. Hence, several studies of VTC emission have been performed already, also for transition metals, where the following assignments of VTC-XES features have been made for compounds of Manganese [8, 85], Chromium [78], Titanium [88] and Iron [51, 67] (schematized in the right panel of Fig. 6.34):

- The $K\beta''$ feature is related to the (1st period) ligand 2s to metal 1s transitions (allowed since different atoms are involved) and therefore is sensitive to the ligands protonation. Thus it allows the identification of the ligand type. Furthermore, for oxygen ligands of Mn, the intensity of the $K\beta''$ feature was shown to be inversely proportional to the metal-oxygen distance.

- The $K\beta_{2,5}$ peak is related to the transitions from p-orbitals (metal 4p and ligand 2p) to metal 1s that do marginally overlap with metal 3d orbitals at the high-energy side, whereby the contributing orbitals are mainly localized at the ligands. For Mn the $K\beta_{2,5}$ peak position was found to be shifted about 1 eV for an unit increase of the formal valency moreover.

In order to check the validity of these assignments for Co, FEFF simulations of angular momentum (quantum number l) projected density of states (l-DOS) simulations of cubic rocksalt CoO-cub, spinel Co_3O_4, Co-carbide Co_2C and $CoCO_3$ were performed (see Table C.3 for crystallographic details). The results are shown in Fig. 6.35 (a) - (d). Here the marginal l-DOS contributions (O-3d, C-3d and Co-4s) are neglected. All features of the two VTC-XES spectra of the references CoO and $CoCO_3$, whose $K\beta_{1,3}$ high-energy tails have been subtracted by fitting them with Voigt functions, are correctly reproduced by FEFF, albeit some features show wrong relative intensities. Thus, the following statements can be made:

- For CoO and Co_3O_4 in Fig. 6.35 (a) and (b) the assignments are identical to those of other transition metals: O(2s)→Co(1s) generates $K\beta''$, and O(2p)→Co(1s) is responsible for $K\beta_{2,5}$ and since Co(4p) do marginally overlap with the O(2p) orbitals, it gives a tiny contribution too. The high-energy shoulder of $K\beta_{2,5}$ above 7710 eV eventually originates from Co(3d)+O(2p)→Co(1s) transitions, i.e. exhibits quadrupole character and is significantly enhanced for the case of Co_3O_4 due to the 4p contribution and the O(2p) increase (both effects originating from the absence of inversion symmetry).

- The Co_2C spectrum in Fig. 6.35 (c) can be equally explained, just that O is replaced by C, and the sensitivity to the ligands protonation shows up. The energetic difference for C(2s)–O(2s) and C(2p)–O(2p) is about 8 eV and 0.5 eV respectively, which implies a clear discriminability of C and O at the $K\beta''$ feature.

- The mixed compound $CoCO_3$ in Fig. 6.35 (d) by contrast, shows a diverse, i.e. more complex l-DOS structure. Here $K\beta''$, mainly arising from O(2s) and split up by C(2s) and C(2p) contributions, is even lower in energy (about 5 eV) than O(2s) in CoO. The first two features of $K\beta_{2,5}$ are due to O(2p) but also due to a mixture of C(2s) and C(2p). The high-energy side lastly is

6. Site-Selective XAS

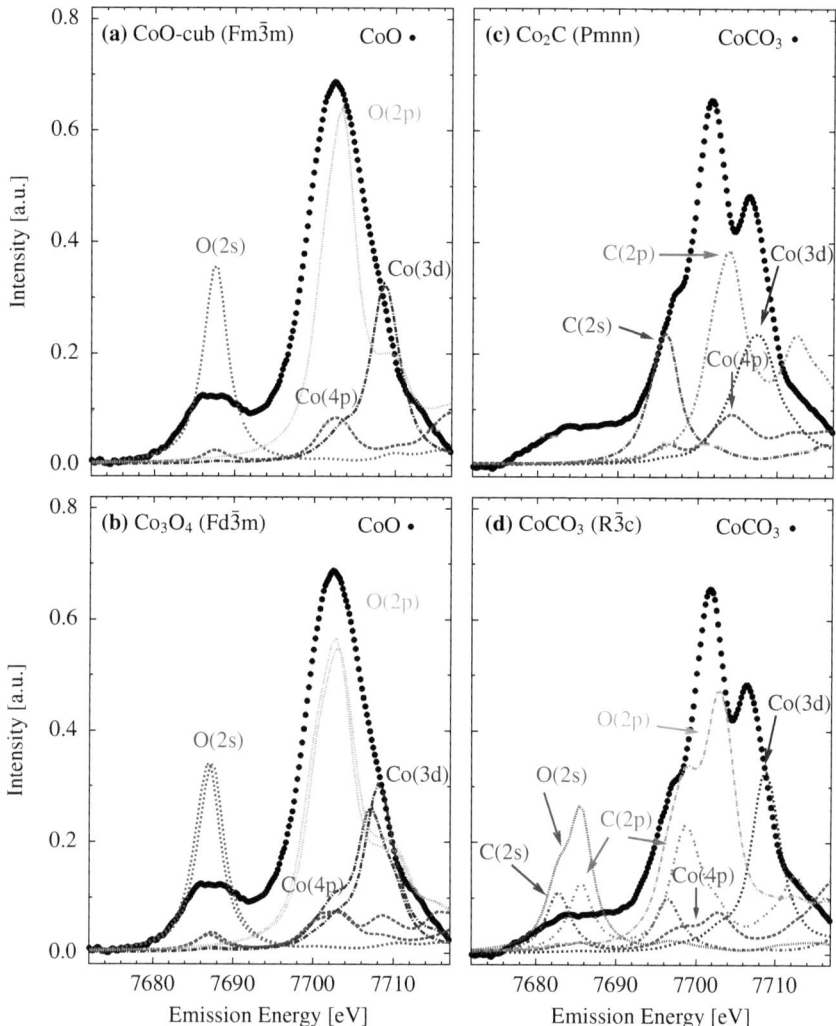

Figure 6.35.: FEFF l-DOS simulations in comparison to experimental VTC-XES spectra (black data points •): (a) rocksalt CoO-cub as well as (b) Co_3O_4 along with CoO on the left and (c) Co_2C as well as (d) $CoCO_3$ along with $CoCO_3$ on the right.

equally composed as in CoO. Consequently, if a Co-compound with different ligands is present in the Co nanoparticles, it will be hard to disentangle the ligands contributions.

Based on these studies, the VTC-XES spectra of the Co nanoparticles will be investigated. The spectra of all samples, after subtraction of the high-energy tail of the $K\beta_{1,3}$ emission line, are shown

6.3. 1s3p-RIXS, HRFD-XANES and VTC-XES at undulator beamline ID26

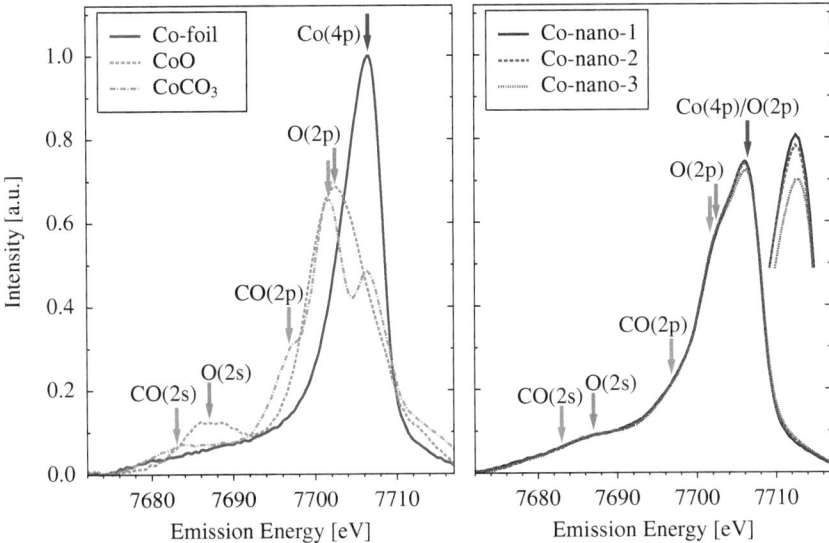

Figure 6.36.: $K\beta_{2,5}$-RXES (or VTC-XES) spectra of the references Co, CoO and CoCO$_3$ (left) as well as of the three Co nanoparticles (right). The arrows indicate the significant features of the references, with labels denoting their origin.

in Fig. 6.36. They are already normalized, since the XES spectra have been normalized with respect to the complete $K\beta$ spectrum (Fig. 6.3.2). The arrows indicate all significant peak positions of the references, with labels referring to their dominant origin. The three Co nanoparticles' spectra are very similar to each other and just show small variations at their respective $K\beta_{2,5}$ peak at 7606.2 eV, which is therefore shown magnified too. All Co-nano spectra exhibit significant contributions from divalent CoCO$_3$ and/or CoO at the two positions of O(2p), visible as a shoulder, and metallic Co at position Co(4p)/O(2p). Furthermore, a small broad $K\beta''$ feature is visible at the positions CO(2s) and O(2s) and significant intensity in between at the position CO(2p).

Table 6.12.: Results of the LCF of the three Co nanoparticles' $K\beta_{2,5}$-XES spectra by Co-foil, CoO and CoCO$_3$. The fit quality is given by the R-factor, and the fitting errors are given in brackets (with respect to last digit).

$K\beta_{2,5}$ LCF	R-factor [e-5]	Ratio (Co : CoO : CoCO$_3$)
Co-nano-1	555	50.1(4) : 37.1(13) : 10.7(13)
Co-nano-2	479	48.9(4) : 39.1(12) : 10.1(13)
Co-nano-3	292	48.3(3) : 34.4(09) : 15.5(10)

The LCF of the nanoparticles' VTC-XES spectra by those of the three references, given in Fig. 6.37 and Table 6.12, reveals that, in contrast to the $K\beta_{1,3}$ emission line, the fit is far more sensitive to the ligands, represented by CoO and CoCO$_3$, and is significantly improved due to inclusion of the

105

6. Site-Selective XAS

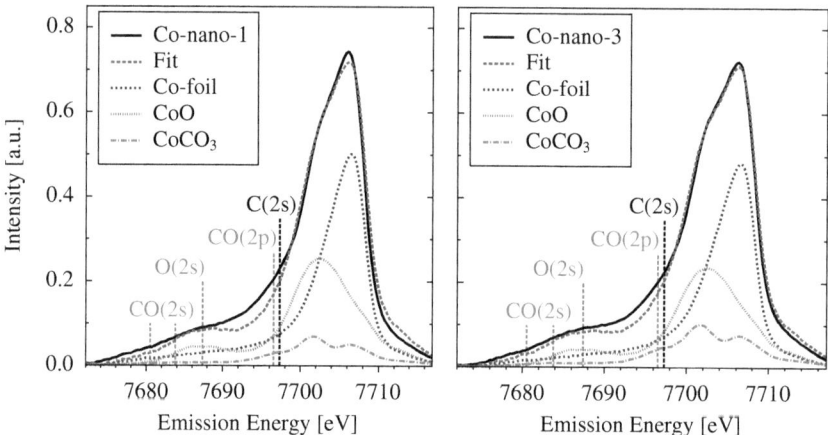

Figure 6.37.: $K\beta_{2,5}$-XES spectra of Co-nano-1 and Co-nano-3 fitted by the references Co-foil, CoO and CoCO$_3$. The reference spectra are scaled with respect to the fitting results. The vertical dashed lines denote the positions of the significant features of CoO and CoCO$_3$ in the low-energy range, as identified by FEFF simulations (see Fig. 6.35).

latter. Furthermore, the explicit ratios shown in Table 6.12 suggest that, within the errors, the valence structure of the three nanoparticles is identical as is expected from the synthesis (section 4.1.1). However, the fit quality is worse about one order of magnitude compared to the $K\beta_{1,3}$ LCF, indicating that some or all of the references are not identical to the real compounds of the Co nanoparticles, which is in accordance with previous LCF's. A closer look shows that in particular the region between $K\beta''$ and $K\beta_{2,5}$ (from 7690 eV to 7698 eV) is significantly underestimated.

With the help of further $K\beta_{2,5}$-XES simulations by FEFF (see Table C.3 for crystallographic details) given in Fig. 6.38 it will be clarified, whether or not all or only some of the fitting components Co-foil, CoO and CoCO$_3$ are appropriate to describe the nanoparticles. The simulations have been convoluted by a Gauss curve of 1 eV FWHM (full width half maximum), to account for the broadening of the spectrometer. However, upon comparing the CoO spectrum (black data points) with its simulation CoO-cub in Fig. 6.38 (b), it is obvious that the latter still shows too sharp features. The energetic positions though, are correctly reproduced, so that the following conclusions can be drawn:

- In Fig. 6.38 (a) the FEFF simulations of the stable metallic Co phases fcc, hcp, and ε are shown along with the spectra of the Co-nano-3 (+ LCF) and Co-foil. Co-foil shows up as being dominantly composed of fcc whose peak positions coincide at 7706.6 eV. The hcp simulation by contrast peaks at 7706.2 eV, perfectly fitting to the Co-nano-3 (and nano-2, 1 likewise) spectrum. Co-ε eventually has a even lower positioned peak at about 7705.8 eV. The nanoparticles' $K\beta_{2,5}$ peak does not allow a clear identification, although the hcp-phase is preferable. Anyway, it can be stated that all differences between fit and experimental data (see Fig. 6.37) at energies below 7704 eV are not due to the metallic core of the nanoparticles, as the Co FEFF simulations are identical there. Consequently, Co-foil is assumed to be an adequate model compound with respect to the VTX-XES region.

6.3. 1s3p-RIXS, HRFD-XANES and VTC-XES at undulator beamline ID26

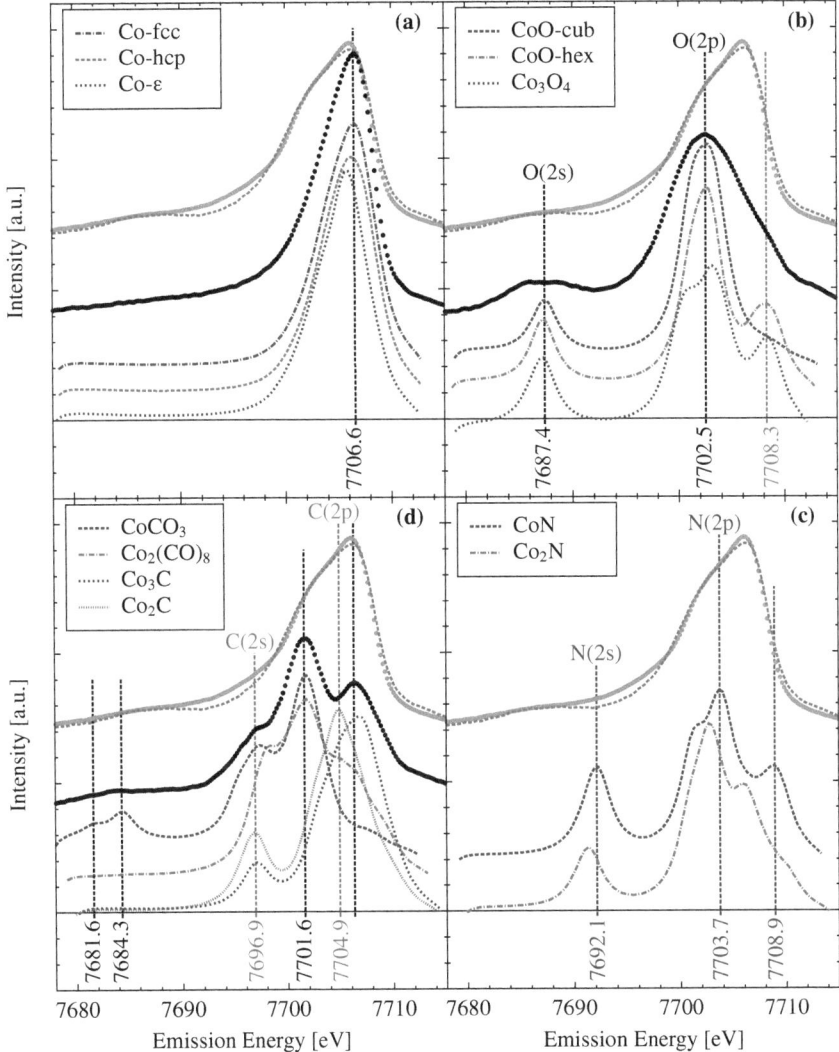

Figure 6.38.: FEFF $K\beta_{2,5}$-XES simulations of (a) Co metals, (b) Co-oxides, (c) Co-nitrogens and (d) Co-carbons, in comparison to Co-nano-3 (gray data points ●) and its LCF (red dashed line - - -) and one of the references (a) Co-foil, (b) CoO and (d) $CoCO_3$ (black data points ●). The vertical lines mark significant features of the references and the simulations, respectively.

- In Fig. 6.38 (b) the FEFF simulations of the stable Co-oxides CoO-cub (cubic Rocksalt), CoO-hex (hexagonal Wurtzite), and the spinel Co_3O_4 are shown along with the spectra of Co-nano-3 (+LCF) and CoO. All simulations almost coincide in the positions of O(2s) at 7687.4 eV and of O(2p) at 7702.5 eV, albeit the main peak of Co_3O_4 is slightly shifted to higher energies, originating from its higher oxidation state. These Co-O features are relevant in the nanoparticles and consequently none of CoO-cub, CoO-hex, or Co_3O_4 can be preferred. At 7708.3 eV, however, the spectra show significant differences, which is due to the breaking of the inversion symmetry (effects mentioned before) in CoO-hex and Co_3O_4. The presence of one or both of the latter two compounds inside the nanoparticles would be possible, if the metal Co contribution were smaller. Anyway, the high-energy region was well reproduced in the LCF, so that finally CoO can be taken as an adequate and preferable fitting compound too.

- The relevant Co-carbon compounds Co_2C, Co_3C, $CoCO_3$, and $Co_2(CO)_8$ (the precursor of the nanoparticles' synthesis, see section 4.1.1) are shown in Fig. 6.38 (c), along with the spectra of Co-nano-3 (+ LCF) and $CoCO_3$. Co-C compounds have significant features at 7696.9 eV due to C(2s) and at 7697.1 eV due to CO(2p) mainly, which both are important for the nanoparticles. Furthermore, $CoCO_3$ is the only compound that shows features below 7685 eV, where the nanoparticles' spectra were also slightly underestimated in the LCF, so that it can be stated an important fitting compound too.

- A possibility to fill the fitting gap from 7690 eV to 7698 eV arises, when considering Co bonded to nitrogen which is abundantly used in the synthesis. The simulations of two standard compounds CoN and Co_2N are shown in Fig. 6.38 (d), along with the spectra of Co-nano-3 (+LCF). In accordance to the stated protonation sensitivity of the ligand 2s level, N(2s) is positioned in between O(2s) and C(2s) at 7692.1 eV for trivalent CoN and somewhat lower for Co_2N due to its lower formal oxidation state. Thus, an appropriate Co-N reference with (probably a) formal oxidation state ≥ 3, would most likely significantly improve the fit in the energy range discussed here.

One can conclude that, just as was shown by the LCF results in Table 6.37, metallic Co, Co ligated with Oxygen and with Carbon, most probably with decreasing amount in that order, is present in the Co nanoparticles. The metallic core is preferable in the hcp phase and the presence of $CoCO_3$ is quite certain, the precise type of the Co-oxide remains open, though rocksalt CoO is preferred. Besides, there are strong indications for the presence of a Co-nitrogen compound, which would lead to different and yet unsettled ratios of the other compounds present in the nanoparticles.

6.3.7. HRFD-XANES from $K\beta_{2,5}$

HRFD-XANES spectra recorded at $K\beta_{2,5}$ fluorescence energies, exhibit site-selectivity due to the underlying valence-to-core (VTC) processes as was shown by the FEFF simulations in Fig. 6.35 and 6.38. One can choose positions

- at the $K\beta''$ feature to select mainly those Co atoms bonded to a particular ligand (L) since the latter contributes with its well separated L(2s) orbital,

- or at the $K\beta_{2,5}$ peak to get at least an increased selectivity to one of the contributing Co-compounds due to the L(2p) orbitals, even to the pure Co metal due to its p orbitals.

6.3. 1s3p-RIXS, HRFD-XANES and VTC-XES at undulator beamline ID26

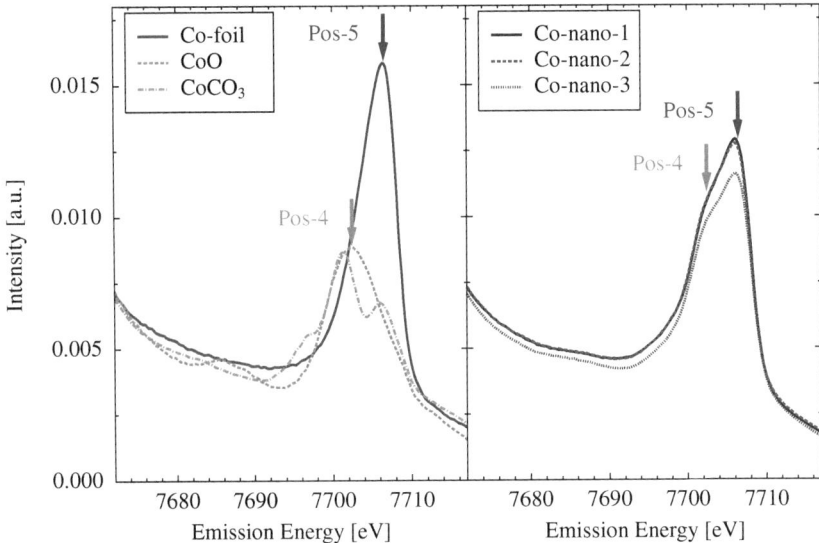

Figure 6.39.: $K\beta_{2,5}$-XES spectra of the three references (left) and the three nanoparticles (right). The spectra are cut off from the normalized $K\beta_{1,3}$ NRXES spectra, whose high-energy tail is visible therefore.

Consequently, site-selectivity is achievable here with respect to the different ligands (ligand-selectivity), independent on the corresponding Co-compounds valency. Due to the low intensity of the VTC process, however, in particular at the $K\beta''$ feature, only two positions were chosen that correspond to the $K\beta_{2,5}$ peaks of CoO at 7702.6 eV (Pos-4) and Co-foil at 7706.6 eV (Pos-5). They are shown in Fig. 6.39, which is similar to Fig. 6.36 where the positions 4 and 5 have been labelled O(2p) and Co(4p). However, now the high-energy tail of the $K\beta_{1,3}$ emission line has not been subtracted, since it is of importance for HRFD-XANES. The 3p→1s transitions, that give rise to $K\beta_{1,3}$, will contribute to the HRFD-XANES spectra recorded at $K\beta_{2,5}$, according to their intensity at the respective positions Pos-4 and 5. An estimation of the contribution of the $K\beta_{1,3}$ high-energy tail to the $K\beta_{2,5}$ peak yields about 23 % for Pos-4 and 18 % for Pos-5. With respect to the uncertainty of this estimation, it will be set to 20 % for both positions. Lastly, since Pos-4 and 5 are energetically just below the onset of the Co K-edge at 7709 eV, their elastic peaks are visible in the respective HRFD-XANES spectra. To avoid disturbance by these peaks, they will be fitted by Voigt functions and then subtracted.

The HRFD-XANES spectra from Pos-4 and Pos-5 for all three nanoparticles are shown in Fig. 6.40. The "noise" in front of the edges is the remnant from the subtraction of the elastic peaks. The complete XANES spectra are more noisy compared to those extracted at $K\beta_{1,3}$ (Pos-1 to 3) due to the small overall intensity of the $K\beta_{2,5}$ fluorescence.

The spectra extracted from Pos-5 are very similar to Co-foil, in particular the one of Co-nano-1 which has the "thickest" core as can be seen by its direct comparison to Co-foil in the left panel of Fig. 6.40 (lower curves). The edge intensity of Co-nano-1 is slightly smaller though, and the whiteline has only one peak at about 7725 eV, indicating the hcp-phase again. Furthermore, the first shape

6. Site-Selective XAS

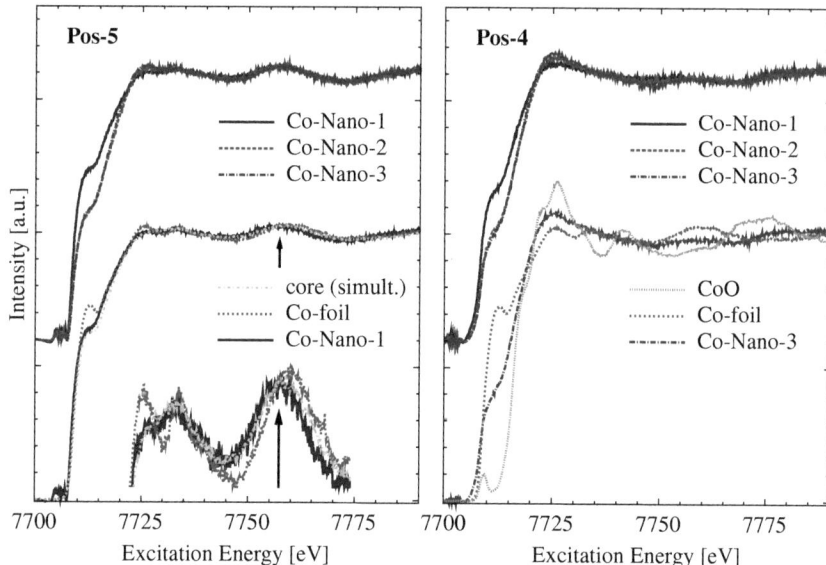

Figure 6.40.: Normalized HRFD-XANES spectra of the three nanoparticles from fluorescence, detected at the $K\beta_{2,5}$ peak positions of CoO in the left panel (Pos-4) and of Co in the right panel (Pos-5). The spectrum of Co-nano-1 is compared to that of Co-foil and the site-specific core (see Fig. 6.22) for Pos-4, and the region from 7722 to 7774 eV is magnified with respect to its intensity. The Co-Nano-3 spectrum from Pos-5 is compared with Co-foil and CoO.

resonance of Co-nano-1 at about 7757 eV (marked by an arrow in Fig. 6.40, see also magnification) is shifted about 2 eV to lower energies compared to Co-foil. This could be due to an increase of the nanoparticles' lattice constant compared to the bulk, which is unlikely, or again due to a bcc-phase contribution. One can already conclude now that the HRFD-XANES spectra from Pos-5 are almost pure site-selective with respect to zerovalent Co. Consequently, the site-specific core spectrum from method S, which was determined in section 6.3.4, is also compared to Co-nano-1 in the left panel of Fig. 6.40. Obviously, both spectra are almost identical, which confirms the pure site-selectivity of Co-nano-1 from Pos-5 on the one hand and the reasonability of the determined site-specific core spectrum on the other hand.

If one takes into account the composition of Pos-5 with respect to the $K\beta_{2,5}$ LCF (see Fig. 6.37), a 69 : 22 : 7 ratio of Co-foil : CoO : CoCO$_3$ (with errors ±0.4 for Co-foil and ±1.3 for the divalent Co's) is expected for Pos-5. However, this has to be corrected by the 20 % $K\beta_{1,3}$ contribution owing to its high-energy tail. Co-nano-1 was calculated to be a 80 : 20 mixture of Co : CoO by the $K\beta_{1,3}$ LCF so that a corrected ratio of about 71 : 27 is yielded, where CoO and CoCO$_3$ are comprised in the 27 %. Thus, up to now Co-nano-1 is not expected to be exclusively a metallic Co spectrum, which implies that either the spectrum of Co-nano-1 from Pos-5 (and that of the site-specific core spectrum likewise) does not reflect a pure metallic Co but some mixture, or the ratio calculated for $K\beta_{2,5}$ is drastically different and the metal Co component is much more dominant. As is known from the last

Table 6.13.: Results for the LCF of the nanoparticles' HRFD-XANES spectra from $K\beta_{2,5}$. The fitting components have been the site-specific core and shell spectrum as determined by method S in section 6.3.4.

HRFD-XANES LCF	R-factor [e-5]	Pos-4	Pos-5
Co-nano-1	36, 35	89.0(4) : 11.0(4)	100.0(0) : 0.0(0)
Co-nano-2	36, 44	70.0(3) : 30.0(3)	85.5(4) : 14.5(4)
Co-nano-3	46, 39	66.9(4) : 33.1(4)	85.7(4) : 14.3(4)

section, the references Co-foil (fcc phase), CoO (rocksalt), and $CoCO_3$ (spinel) were not sufficient to describe the nanoparticles $K\beta_{2,5}$-XES spectra. Consequently, upon performing a LCF, with the as yet not precisely specified real Co-compounds of the nanoparticles, the resulting ratio of the metallic Co could be much higher. The most important point, however, is that the HRFD-XANES spectra from $K\beta_{2,5}$ do not exhibit the same sensitivity to the reference compounds compared to the $K\beta_{2,5}$-XES spectra. Since, although the orbitals relevant for $K\beta_{2,5}$ are located at the ligands mainly, the Co p-Orbitals are overlapping with them at almost each significant feature (see Fig. 6.35), and as was shown by DFT calculations [43, 8, 51] the mayor oscillator strength of the transition stems from these Co contributions. As a consequence, HRFD-XANES from $K\beta_{2,5}$ exhibits a significantly smaller sensitivity to the ligands compared to XES at $K\beta_{2,5}$.

Based on these discussions, the conclusion can be drawn that both the HRFD-XANES spectrum of Co-nano-1 from Pos-5 of $K\beta_{2,5}$ and the core spectrum determined by the SVD of the HRFD-XANES spectra from $K\beta_{1,3}$, represent the pure site-selective core of the nanoparticles. The result is of course afflicted by the previously discussed uncertainties due to lifetime-influences as well as fitting errors.

The spectra extracted from Pos-4 in the right panel of Fig. 6.40 (upper curves) all show a lower edge and more intense whiteline compared to those from Pos-5, i.e. they show an enhanced Co^{2+} and reduced Co^0 contribution. This can be recognized clearly upon comparison of the spectrum of Co-nano-3 (which has the "thickest" shell) with those of Co-foil and CoO in the lower curves of the same figure. The Co^{2+} amount is increasing from Co-nano-1 to 3, consistent with the synthesis (section 4.1.1) and the results from section 6.3.4. However, a reliable LCF of these Co-nano spectra with experimental data can not be performed, as no references have been measured at Pos-4 and 5. Instead, motivated by the result for Co-nano-1 at Pos-5, a LCF of the HRFD-XANES spectra by the site-specific core and shell spectra is performed with results shown in Table 6.13.

According to the quality of the LCF it is clear that the utilized site-specific core and shell spectra can be easily reproduced via SVD applied to the HRFD-XANES spectra from $K\beta_{2,5}$. The more important task would be to separate the shell spectrum into several parts, describing Co compounds with the dominant ligands O, N and C, as was tried for the HRFD-XANES spectra from $K\beta_{1,3}$ (see Fig. 6.26). A respective SVD, however, failed already when searching for two shell components upon conservation of the core ratios (Table 6.13): either the spectra became unphysical or far too noisy to be interpretable. Actually it was shown before (see Fig. 6.26) that searching for three, instead of two, shell components yielded more reasonable results. However, due to the worse quality of the spectra this was not possible in the current case.

All in all, site-selectivity (or ligand-selectivity) via HRFD-XANES measurements from $K\beta_{2,5}$ fluorescence showed to be a promising tool. Without any further data processing, the pure site-selective core spectrum for the Co nanoparticles could be obtained upon recording the fluorescence from the $K\beta_{2,5}$ peak of metallic Co. With sufficient photon flux as well as huge integration times, site-selective spectra of the Co-ligand compounds should be achievable upon recording HRFD-XANES spectra

6. Site-Selective XAS

from the fluorescence of the $K\beta''$ features.

6.3.8. Summary of Co nanoparticle properties

With the help of the various measurements performed at ESRF's ID26 beamline and the techniques applied, the Co nanoparticles could be thoroughly characterized. No contradictions to the initially presumed simple core-shell model and its implications have been encountered. The EXAFS results in particular confirmed the onion-like core-shell structure and for the two sites, core and shell, the following information could be gathered:

- The metallic core, with diameter between 46 and 53 Å (depending on the synthesis, see section 4.1.1), is present in the Co-hcp phase dominantly (with minor Co-fcc phase contribution probable) and, regarding the local structure, is highly crystalline with lattice constants $a = 2.489(6)$ Å and $c = 4.065(18)$ Å.

- The shell's thickness amounts between 3.5 and 7.0 Å (depending on the synthesis, see section 4.1.1) allowing for 1 – 3 monolayers only. The presence of C, N and O is certain. Hereby the dominant Co compounds have a valency of +2, most probably adopted by CoO (cubic and/or hexagonal) and $CoCO_3$. The Co-N compound should exhibit a valency of at least +3. These first period ligands are located at 2.00(1) Å around the Co atoms, which themselves are separated about 2.99(2) Å from each other.

Furthermore, it was found that the shell thickness, within the available range, has no visible effect onto the core's properties accessible by the utilized X-ray techniques.

6.4. Conclusion

The strategy for achieving site-selective XANES spectra, by means of a numerical procedure applied to deliberately chosen distinct HRFD-XAS spectra, was first applied to a 80 : 20 Mix of Co and CoO in section 6.2.1. It was partly successful, as only the dominant Co component could be extracted from the HRFD-XANES spectra, but not CoO, most probably due to its small contribution in the Mix and due to the insufficient resolution and X-ray flux of the measurements. Hereby, however, a sensitivity limit of the numerical procedure was found and in the next section 6.2.2, where a "real" system of Co nanoparticles with estimated ratio of Co-metal : Co-oxide of 56 : 44 have been investigated, two apparently reasonable pure site-selective components (core and shell) were found. The site-selective (mainly divalent) shell resembled fully oxidized Co nanoparticles and the (metallic) core a mixture of Co-ε and Co-hcp, with respect to the electronic and geometric structure. However, both the core and the shell could not be determined uniquely, but with some variability, which was due to the still too low resolution of these measurements. Moreover, as a result of the relatively low resolution, the core was determined partially incorrect, which became apparent in the next series of measurements.

These last measurements, in section 6.3, were carried out on three types of Co nanoparticles with estimated ratios of 80 : 20, 73 : 27 and 58 : 42 (Co-metal : Co-oxide) and with high resolution, sufficient finally to determine unique site-selective spectra: a Co-hcp core and a shell of divalent CoO and $CoCO_3$ as well as a (unspecified) Co-N contribution. Noteworthy is that, in contrast to the first study, the "80 : 20" nanoparticles yielded reasonable site-selective spectra, even for the weakly present component. Owing to the three types of nanoparticles, three methods for the determination of pure site-selective XAS spectra became available: One set of site-selective spectra for each type

6.4. Conclusion

of nanoparticle (method 1), one set for each position where the HRFD-XAS spectra were recorded (method 2) and lastly one set for all HRFD-XAS spectra simultaneously (method S). Hereby method 1 was the one applied in the previous sections too and which failed when searching for site-selective EXAFS spectra, probably due too lack of statistics. Method 2 by contrast would have yield site-selectivity on the basis of differences in the HRFD-XAS spectra due to the synthesis alone, however, it failed for XANES and EXAFS likewise. The reason for this was the non-uniform distribution of the sought site-selective components in the respective sets of HRFD-XAS spectra, which conflicts with the previously found sensitivity limit. Finally, Method S yielded, in all cases, the most reasonable and reliable site-selective spectra. As a consequence, it must be stressed that it is highly recommended to have a redundant number of HRFD-XAS spectra, i.e. when for example searching for two site-selective components, neither two nor three HRFD-XAS spectra, but e.g. six for XANES or even nine for EXAFS are adequate. This redundant number of spectra, however, should not be gained (only) by choosing many positions for HRFD-XAS recording (this failed for the Mix measurements), but also with another "method" (e.g. synthesis) that allows the same variation as achieved by HRFD spectroscopy.

In order to get reasonable site-selective spectra in general, it was found moreover that both for XANES and EXAFS (using method S) not the precise ratios of the linear combination fits could (and neither must) be reproduced, but the trends. This was an important point as the references, used to describe and fit the nanoparticles, had to be chosen initially, without certainly knowing their suitability – and which is even more important for completely unspecified materials. Nonetheless, for XANES in particular, suitable references are most important to interpret the final site-selective spectra and the lack of those in the current work, had been the reason for the limited description of the nanoparticles' shell. Another important point was the disturbance of the HRFD spectra in the case of XANES due to lifetime-broadenings. It was found that by choosing the positions for HRFD-XAS appropriately – e.g., for three positions: one close to the emission peak and the other two equidistantly at lower and higher energy – these disturbances were almost negligible (as they cancel out).

All in all, site-selectivity on the basis of valency-selectivity is an advanced technique to disentangle the different site-specific components of a multi-valency compound. If all mentioned conditions can be fulfilled and if appropriate references and simulations are available, the determination and identification of the site-selective spectra is possible. No other technique available currently, is capable of providing this kind of information just via one experiment.

7. Summary and Outlook

In this work, thorough investigations of metal nanoparticles by X-ray absorption and X-ray emission spectroscopic (XAS/XES) techniques as well as a combination of these, high-resolution fluorescence-detected XAS (HRFD-XAS), have been presented. It has become evident that XAS is a convenient tool for nanoscaled systems as it gives direct insight into the electronic valence structure as well as the geometry, via the X-ray absorption near edge structure (XANES), and to the local atomic arrangements, via the extended X-ray absorption fine structure (EXAFS), independently on crystallinity or size of the probed material.

For the special case of tempered Co_3Pt/C nanoparticles in chapter 5 the Pt L3-edge XANES revealed a lowered Pt 5d band center, which was found by EXAFS analysis to be mainly due to a tightened Pt–Pt bond. As a consequence, oxygen is less tightly bound when adsorbed to the Pt surface, which led to a significant increase of the oxygen reduction reaction (ORR) activity when Co_3Pt/C is applied as cathode material in a PEMFC fuel cell. As was suggested by [111], one can also expect an alleviation of the poisoning by carbon-monoxide, which results from impure hydrogen fuel, as a result of the decreased adsorption strength. With the help of EXAFS, the reason for the short Pt–Pt distance could be attributed to a strain effect due to the high fraction of Co atoms present in the (highly) disordered PtCo-fcc lattice. An uncertainty, however, remains regarding these results as XAS spectra reflect the average of all (Pt) atoms present in the particle, those at the surface (layers) which are most relevant for the catalysis and those in the bulk that play a secondary role only and may have different properties. Here, surface sensitive techniques like X-ray photoelectron spectroscopy (XPS) or site-selective XAS would have complemented this study.

Nonetheless, the obtained results suggest that further improvement of the Co_3Pt/C catalyst could be achieved upon increasing the amount of alloyed Co. However, as it was found that unalloyed Co is still present in the catalyst, it is not a matter of increasing the amount of the initial Co precursor but of a suitable post-synthesis treatment, e.g., prolonging of the tempering process at about 800 °C might enforce the alloying, since lower and higher temperatures showed significantly lower ORR activities. Here moreover, some restrictions had to be considered: There is a lower limit of the Pt–Pt bond distance where the adsorption of oxygen becomes too loose to be applicable for the ORR. Further on, the increased fraction of alloyed Co also leads to a depletion of the surface Pt atoms and at some point the ORR activity could decrease as a result, unless surface Co atoms get lost by annealing or during fuel cell operation. As a result of both effects, the surface could even become catalytically "inert". Lastly, it is probable that the alloy phases present in Co_3Pt/C will change with increasing Co fraction but, as was suggested by X-ray diffraction measurements [81], they are crucial for the total ORR activity too. To sum up, the Co_3Pt/C nanoparticles are promising candidates to improve the efficiency of fuel cells and further on, are still capable of being further optimized with respect to their catalytic properties.

The central focus of this thesis was the establishment of site-selective XAS for nanoparticles (and mixed-valency compounds in general), which implies a significant technical improvement for XANES/EXAFS as it in principle allows to describe "interior" (bulk) and "surface" (coating) of a nanoparticle separately. It was successfully tested on a system of Co nanoparticles, as discribed in chapter 6. There, the explicit phase of the metallic crystalline interior (the "core") could be deter-

7. Summary and Outlook

mined to be dominantly hexagonal-close-packed (hcp), with lattice constants slightly smaller than that of bulk Co. For the surface or rather coating (called "shell") no crystallographic phase was found, most probably as it is too thin (only a few layers) to develop one. However, its valency and interatomic distances were accessible: A divalent shell of mainly Co-oxide and Co-carbonate, with average Co–Co distance similar to a rocksalt Co-oxide, i.e. (2.99 ± 0.02)Å and Co–ligand distance rather short, only (2.00 ± 0.01)Å. The following prerequisites were identified to be important in order to get meaningful results in general:

1. The lifetime broadening influences onto XANES, as explained in section 2.3.3, turned out to be of no importance, if the positions for HRFD-XAS recording were chosen on the fluorescence peak as well as equidistantly to lower and higher energy relative to it. The distortions of the XANES spectra are just in opposite directions then, so that they cancel out each other, for example, the edge-onset is once shifted to lower and once to higher energy by the same amount.

2. There is (of course) a sensitivity limit with respect to the fractions of the sites, which depends on the data quality, i.e. on the quality of the experimental station and the flux provided by the synchrotron. In order to check this, it is advised to perform measurements on a "test system" (physical mix of two chemically different species of an element) first that reflects the ratios present in the actual material of interest.

3. For the case of significantly unequal fractions of the different components, moreover, it is recommended to have samples with other ratios at hand as well, for example, for two components at best with opposite and equal ratio to avoid an (over-) underestimation of the (strong) weak component. Hereby, the statistics of the numerical procedure (SVD) are improved, and actually it is recommended to have generally a redundant number of experimental HRFD-XAS spectra with differing ratios that, moreover, are not only obtained by choosing different fluorescence positions but, e.g., by variations in the synthesis.

In summary, the main goal was achieved: The establishment of site-selective XAS for nanoparticles. The secondary goal, however, the identification of a bulk–coating interaction could not be accomplished as the Co nanoparticle system was inadequate for this purpose. Its coating turned out to have no significant influence onto its core, at least not from the viewpoint of XAS/XES and for the shell thicknesses available. However, it had become obvious that the elaborated strategy for site-selective XAS is advantageous for all types of nanoparticles with weakly and strongly interacting coatings (and for mixed-valency compounds in general), as well as for the complete surface science community. Here especially the field of catalysis could benefit from site-selective XAS, as catalytically relevant interconnections inside the surface as well as between surface and bulk could be disentangled. Here in particular the local atomic arrangements in the surface, provided by site-selective EXAFS, are barely accessible via another technique.

Appendices

A. Simple Models

Surface to Bulk Ratio

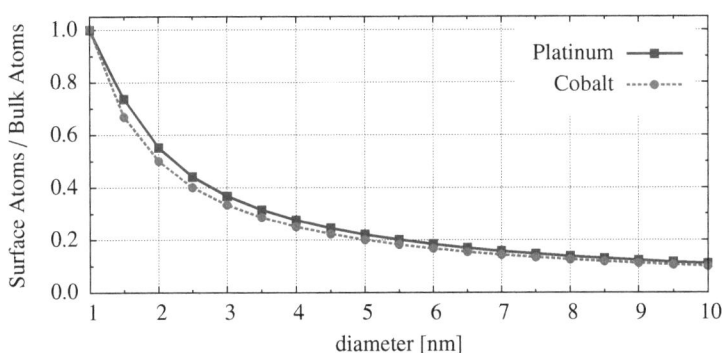

Figure A.1.: Ratio of surface atoms to bulk atoms according to Eq. (A.5) for Pt and Co on a linear scale.

In order to make estimations regarding the surface atoms to bulk atoms ratio, a simple geometrical model is considered. For particles with known size, the volume V and surface S can be calculated, e.g. assuming spheres, according to

$$V_{sphere} = 4/3\,\pi R^3, \quad (A.1)$$
$$S_{sphere} = 4\,\pi R^2, \quad (A.2)$$

with the radius R of the assumed sphere. Based on the known radius R_a of an atom, e.g. 1.253 Å for Co or 1.38 Å for Pt, the volume V_a and geometrical cross section A_a of that atom can be calculated as well:

$$V_a = 4/3\,\pi R_a^3, \quad (A.3)$$
$$A_a = 2\,\pi R_a^2. \quad (A.4)$$

The resulting number of atoms for the surface divided by those of the bulk gives the desired ratio:

$$R = \frac{A_a/S_{sphere}}{V_a/V_{sphere}}. \quad (A.5)$$

It is shown in Fig. A.1 for Co and Pt and is reasonable down to diameters of about 2 nm. For smaller particle sizes more sophisticated calculations are mandatory to make correct predictions, as can be

A. Simple Models

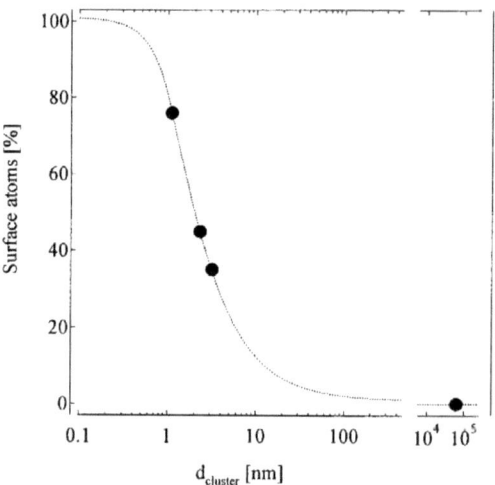

Figure A.2.: Ratio of surface atoms to bulk atoms for Palladium, taken from [63] with logarithmic scale on the x-axis.

seen in Fig. A.2, which is taken from Ref. [63]. There the ratio R for Palladium (Pd) was determined to 35 %, 45 % and 76 % for nanoclusters with diameter of 3.37 nm, 2.5 nm and 1.2 nm. Calculations according to Eq. (A.5) for Pd yield, 32 %, 44 % and 92 %, i.e. almost identical values, except for the smallest cluster of 1.2 nm.

Core-Shell Sizes

A simple core-shell model is considered, based just on spherical geometry, that allows to estimate the thickness of the core and shell of a particle relative to its size $D = 2 \cdot R$, for given atomic ratio abd densities of the core and shell species. The volume of a sphere can be calculated according to

$$V = 4/3 \pi R^3 \quad (A.6)$$
$$= m/\rho \quad (A.7)$$
$$= nM/\rho \quad (A.8)$$
$$= N/N_A M/\rho \quad (A.9)$$

with the density ρ in [g/cm^3], the mass m in [g], the amount of a substance n in [mol], the molar mass M in [g/mol], as well as the number of atoms N and the Avogadro constant $N_A = 6.02214129(27) \times 10^{23}$ mol^{-1}. Assuming a spherical particle with well separated core and shell of radius R_c and $R_s = R - R_c$, the respective volumes can be calculated via

$$V_c = N_c/N_A M_c/\rho_c = 4/3 \pi R_c^3, \quad (A.10)$$
$$V_s = N_s/N_A M_s/\rho_s = 4/3 \pi (R^3 - R_c^3). \quad (A.11)$$

The core radius R_c relative to the total particles' radius R consequently can be obtained:

$$\frac{V_c}{V} = \frac{V_c}{V_c + V_s} = \frac{1}{1 + V_s/V_c} = \frac{R_c^3}{R^3}$$

$$\Leftrightarrow \frac{R_c}{R} = \frac{1}{\sqrt[3]{1 + V_s/V_c}}, \tag{A.12}$$

with V_s/V_c according to Eqs. (A.10) and (A.11):

$$\frac{V_s}{V_c} = \frac{N_s M_s / \rho_s}{N_c M_c / \rho_c}. \tag{A.13}$$

R_s can be calculated analogue to Eq. (A.12):

$$\frac{R_s}{R} = \frac{1}{\sqrt[3]{1 + V_c/V_s}} = 1 - \frac{R_c}{R}. \tag{A.14}$$

For the case of a core-shell particle with different elements (or species thereof) "1" (in the core) and "2" (in the shell), the ratio of shell atoms to core atoms N_s/N_c in Eq. (A.13) can be obtained from a linear combination fit (LCF) of a XAS or XES spectrum. The element ratio N_2/N_1 resulting from the LCF is equal to N_s/N_c, since the spectral intensity in XAS/XES is proportional to the number of absorbers (= atoms) in the sample.

B. Singular Value Decomposition

"In linear algebra, the singular value decomposition (SVD) is a factorization of a real or complex matrix, with many useful applications in signal processing and statistics. Formally, the singular value decomposition of an $m \times n$ real or complex matrix M is a factorization of the form

$$M = U\Sigma V^\dagger, \tag{B.1}$$

where U is an $m \times m$ real or complex unitary matrix, Σ is an $m \times n$ diagonal matrix with non negative real numbers on the diagonal, and $V^\dagger$ (the conjugate transpose of V) is an $n \times n$ real or complex unitary matrix. The diagonal entries $\Sigma_{i,i}$ of Σ are known as the singular values of M. The m columns of U and the n columns of V are called the left singular vectors and right singular vectors of M, respectively.

Singular value decomposition and eigendecomposition are closely related. Namely:

- The left singular vectors of M are eigenvectors of $MM^\dagger$.
- The right singular vectors of M are eigenvectors of $M^\dagger M$.
- The non-zero singular values of Σ are the square roots of the non-zero eigenvalues of $MM^\dagger$ or $M^\dagger M$.

Applications which employ the SVD include computing the pseudoinverse, least squares fitting (LSF) of data, matrix approximation, and determining the rank, range and null space of a matrix." [106]

In this work the SVD will be employed to perform a LSF of a system of inhomogeneous linear equations

$$S_{exp}^i = \sum_k S_{th}^k\, c_{th}^{i,k}, \tag{B.2}$$

whereas the S_{exp}^i will be a number i of experimental absorption spectra that are assumed to be linear combinations of a limited set of "theoretical" spectra S_{th}^k. Here, the S_{th}^k are constraint to be "physical" spectra too, i.e. they have to be positive "functions" of the energy. The $c_{th}^{i,k}$ give the ratios of the theoretical spectra and fulfill the condition $\sum_k c_{th}^{i,k} = 1$, $\forall i$. Eq. (B.2) can be rewritten in matrix notation as

$$M_{exp} = M_{th} C_{th}. \tag{B.3}$$

Here the rows of M_{exp} are the S_{exp}^i, the rows of M_{th} the S_{th}^k and the columns of C_{th} are the $c_{th}^{i,k}$. Both the theoretical spectra S_{th}^k and the coefficients $c_{th}^{i,k}$ are to be determined in the LSF. In other words the aim is to minimize the two-norm $\|M_{exp} - M_{th} C_{th}\|$.

According to the SVD, the matrix M_{exp} can be factorized into two orthogonal matrices U and V and a diagonal matrix L, containing the singular values (SV) in decreasing order:

$$M_{exp} = U\, L\, V^T. \tag{B.4}$$

These SVs also give the importance of the sought spectra, dividing them into "real" and "noise" spectra. Switching back to the physical point of view, $U =: M_{th}$ is comprised of the calculated pure

B. Singular Value Decomposition

spectra and $LV^T =: C_{th}$ represents the corresponding calculated coefficients. Of importance now is that the orthogonal matrices U and V are non-unique, which can be expressed via the invertible square matrix T:

$$M_{exp} = M_{th}TT^{-1}C_{th}. \qquad (B.5)$$

This implies that each theoretical spectrum can be varied independently of each other under simultaneous variation of its corresponding coefficient. One then ends up with a set of mathematical solutions to Eq. (B.5), $M'_{th} = M_{th}T$ and $C'_{th} = T^{-1}C_{th}$, each of which being an equally good result of the LSF and from whom the unique physical solutions, i.e. real spectra, have to be separated. The SVD will be performed with the help of the FITIT software [84] where it is implemented as described here, and the matrix entries of T are accounted for as parameters.

C. XANES

XAS normalization

All experimental XAS data $\mu(E)$ is "normalized" prior to its investigation. The normalization procedure is performed with respect to the complete XAS spectrum in energy space, however, the effect is only "visible" for XANES, since EXAFS spectra are anon transformed to k and/or R-space. "Normalization is the process of regularizing the data with respect to variations in sample preparation, sample thickness, absorber concentration, detector and amplifier settings, and any other aspects of the measurement." [68] The normalized data thus becomes independent on experimental details what makes it comparable with other data and in particular with theory. The normalization process is performed

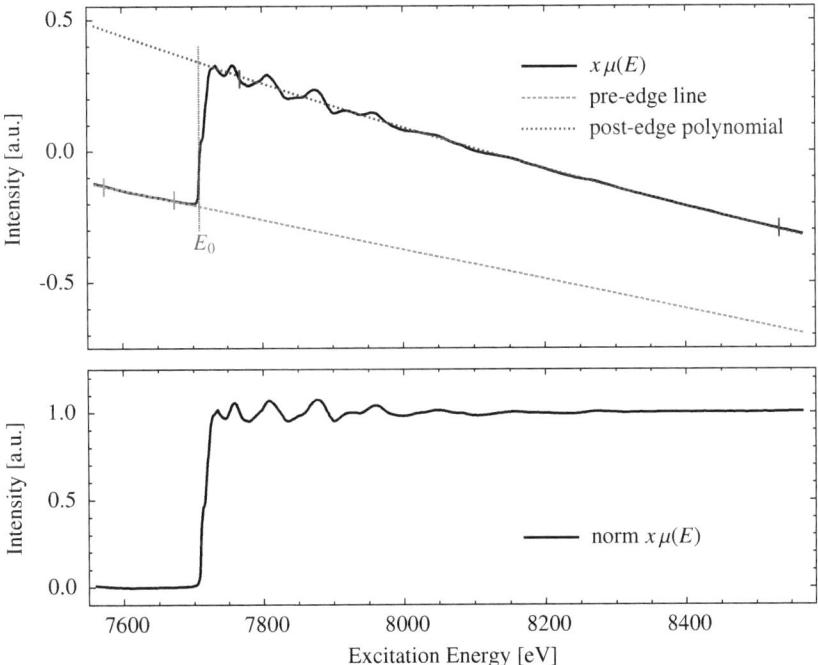

Figure C.1.: Raw experimental XAS spectrum for a Co powder measured in transmission (top), along with pre-edge line and post-edge polynomial that are used to determine the normalized spectrum at the bottom.

C. XANES

by the software Athena [69] as shown in Fig. C.1 and explained step-by-step in the following:

- Linear regression of the pre-edge region of $\mu(E)$. The condition here is to set the starting and ending point to a flat region where no features are visible, like an elastic-peak or even the pre-edge due to the first electron transitions.

- Regression of the post-edge region of $\mu(E)$ by a 2nd-order polynomial. Care has to be taken about the position of the starting point, which must not be on the whiteline or another local maximum, nor oppositely on a local minimum, but somewhere in the middle, so that the fitted polynomial goes through the "middle" of the EXAFS oscillations.

- Both regression functions are extrapolated beyond the edge position defined by E_0. The intensity difference between the pre- and post-edge line at the position of E_0 (the edge or binding energy) results in the so-called edge-jump $\mu_0(E_0)$.

- The pre edge line then is further extrapolated to all energies of the measurement range and subtracted from the experimental data. Afterwards the data is divided by $\mu_0(E_0)$.

- To increase the comparability even more, a so-called flattening procedure is performed that just pushes the post-edge data up to a constant line at intensity = 1.

The effect of this normalization can be seen in Fig. C.1 for a Co powder, where the raw experimental spectrum (top) is given in comparison to the normalized one (bottom). The interval for both the pre-edge and post-edge regression, indicated by small vertical bars, and the edge energy E_0 is also shown. For a more detailed description of the normalization see the documentation of the Athena software by B. Ravel [68].

XANES simulations

XANES spectra are simulated by the FEFF9 [75] software as well as by FDMNES [42] that both are based upon the multiple scattering formalism as introduced in section 2.2.2, albeit FDMNES has a relativistic extension based on the finite-difference method. A typical FEFF9 input file for the calculation of Co K-edge XANES spectra is given in Table C.1 (see [70] for complete documentation of FEFF9 control cards). The following options have been used there:

- The "random phase approximation" (RPA) for the core-hole, first introduced by D. Bohm and D. Pines [11] and strictly proven later by M. Gell-Mann and K. Brueckner [23].

- Self-consistent potential calculations (SCF) (instead of non-self-consistent overlapped atomic potential).

- The "full multiple-scattering" (FMS) procedure (see Eq. 2.22 or Eq. 2.27).

- Usage of Dirac-Hara (DH) exchange-correlation potential plus an imaginary Hedin-Lundqvist (HL) part (the latter could be utilized alone instead of Dirac-Hara).

FDMNES Co K-edge XANES simulations have been calculated likewise FEFF simulations, except for the following differences:

Table C.1.: Standard FEFF9 input file as utilized for all XANES simulations in this work.

EDGE	K					
S02	1.0					
COREHOLE	rpa					
*	pot	xsph	fms	paths	genfmt	ff2chi
CONTROL	1	1	1	1	1	1
PRINT	1	0	0	0	0	0
*	rscf	[lscf]	[scf]	[ca]	[nmix]	
SCF	5.50	0	30	0.2	1	
*	ixc	vr0	vi0	[ixc0]		
EXCHANGE	3	0	0	2		
*	kmax	[kstep]	[Estep]			
XANES	8.0	0.07	0.3			
*	rfms	[lfms]				
FMS	7.0	1				
POTENTIALS						
*	ipot	Z	[tag]	[lmax1]	[lmax2]	[xnatph]
	0	27	Co	-1	-1	0
	1	27	Co	-1	-1	1
ATOMS						
	(list of xyz coordinates of atom-cluster)					

- "Final state rule" (FSR) instead of RPA for treatment of the core-hole. FSR calculates the potentials in the presence of a fully relaxed (screened) core-hole, whereby core-hole – electron interactions (excitons, excitonic resonances, etc.) that are in most cases negligible, are excluded [95]).

- Non-SCF overlapped potentials.

- Real HL correlation exchange potential.

FDMNES calculations with SCF potentials as well as upon utilizing the relativistic option of the "Finite Difference Method", where the shape of the potential is free (no muffin-tin simplification), have been tested too. However, for the Co compounds no significant improvements were achieved.

C. XANES

Table C.2.: LCFs of the three metallic Co K-edge XANES spectra by FEFF9 simulations of Co-fcc and Co-hcp (computational details in Table C.1).

	fcc : hcp	error	R-factor [e-5]
Co-foil (ESRF)	47 : 53	±3	123
Co-powder (ANKA)	9 : 91	±3	16
Co-foil (ANKA)	0 : 100	±4	35

As the Fermi energy is systematically miscalculated by MS-theory, the FEFF9 (and FDMNES) simulations have to be shifted in energy. To find the proper shift for the metal Co simulations, the Co-fcc spectrum has been aligned to the first whiteline peak at 7725.6 eV of Co-foil (HRFD-XANES spectrum measured at the ESRF, see section 3.4), which leads to an agreement of the subsequent features as well, as can be seen in Fig. C.2 (a). An exception here is the pre-edge, however, whose position could not correctly be reproduced. The FEFF9 Co-fcc (and Co-hcp) spectrum have also been simulated using of the HL exchange potential (see dotted lines at the bottom of Fig. C.2) though by worsening of the overall agreement between theory and experimental data, especially with respect to the energetic positions of the features. The alignment eventually yields a shift for Co-fcc of $\Delta E = -15$ eV is then likewise applied to all the other simulated metallic Co spectra. Furthermore linear combination fits (LCFs) of this ESRF Co-foil spectrum, and of Co-foil as well as Co-powder measured at ANKA (see section 3.2), by the FEFF9 simulations of Co-fcc and Co-hcp are also given in Fig. C.2 (a). The fit details are given in Table C.2. For FDMNES simulations of the metal Co phases a shift of $\Delta E = +7.2$ eV was necessary to align the whiteline of Co-fcc to Co-foil (ESRF).

The FEFF9 simulations of the Co-oxides, CoO-cub (rocksalt CoO) and Co_3O_4 (diamond phase) measured as powders at ANKA (see section 3.2) are aligned with their respective references. Here the focus have been onto the whitelines which resulted in $\Delta E = -9$ eV for both Co-oxides. This is shown in Fig. C.2 (b) and gives a good agreement of experimental data with theory for CoO, but not for Co_3O_4, where the simulated shape resonances are too high in energy. The pre-edges are also miscalculated at too low energies like for the Co metals. The FDMNES simulations are likewise aligned and the energy shift is $\Delta E = +7.0$ eV for CoO and $\Delta E = +6.1$ eV for Co_3O_4.

All crystallographic details of the simulations shown here and in this work are listed in Table C.3.

Table C.3.: Crystallographic data of all simulated Co compounds.

	crystal system	spacegroup (synonym)	Schoenflies	lattice constants [Å]			atomic positions [dimensionless]		Ref.
Co-fcc	cubic	$Fm\bar{3}m$	O_h^5	3.5441	3.5441	3.5441	Co	(0;0;0)	[76, 31]
Co-hcp	hexagonal	$P6_3/mmc$	D_{6h}^4	2.5071	2.5071	4.0695	Co	(1/3;1/3;1/4)	
Co-bcc	cubic	$Im\bar{3}m$	O_h^9	2.8270	2.8270	2.8270	Co	(0;0;0)	[87, 22]
Co-ε	cubic	$P4_132$ (β-Mn)	O^7	6.0970	6.0970	6.0970	Co	(v;v;v), (1/8;u;u+1/4)^a	
CoO-cub	cubic	$Fm\bar{3}m$ (rocksalt)	O_h^5	4.2667	4.2667	4.2667	Co	(0;0;0)	[82]
							O	(1/2;1/2;1/2)	
CoO-hex	hexagonal	$P6_3mc$ (wurtzite)	C_{6v}^4	3.2518	3.2518	5.1967	Co	(1/3;1/3;0)	[83, 113]
							O	(1/3;1/3;0.345)	
Co$_3$O$_4$	cubic	$Fd\bar{3}m$ (diamond)	O_h^7	8.0837	8.0837	8.0837	Co	(1/8;1/8;1/8), (1/2;1/2;1/2)	
							O	(u,u,u)^b	
CoCO$_3$	trigonal	$R\bar{3}c$	D_{3d}^6	4.6618	4.6618	14.9630	Co	(0;0;0)	[18, 58]
				(hexagonal-axes)			C	(0;0;1/4)	
							O	(0.2766;0;1/4)	
Co$_2$C	orthorhombic	Pmnn	D_{2h}^{12}	2.8969	4.4465	4.3707	Co	(0;0.347;0.258)	[58]
							C	(0;0;0)	
Co$_3$C	orthorhombic	Pmma	D_{2h}^{16}	4.5160	5.0770	6.7270	Co	(1/3;0.183;0.065), (5/6;0.40;1/4)	
							C	(0.470;0.860;1/4)	
CoN	cubic	$F\bar{4}3m$ (zincblende)	T_d^2	4.2970	4.2970	4.2970	Co	(0;0;0)	
							N	(1/4;1/4;1/4)	
Co$_2$N	orthorhombic	Pmnn	D_{2h}^{12}	2.8535	3.3443	4.6056	Co	(0;0.261;0.325)	
							N	(0;0;0)	

a ($v = 0.06361$, $u = 0.20224$), b (oxygen parameter $u = 0.263$)

C. XANES

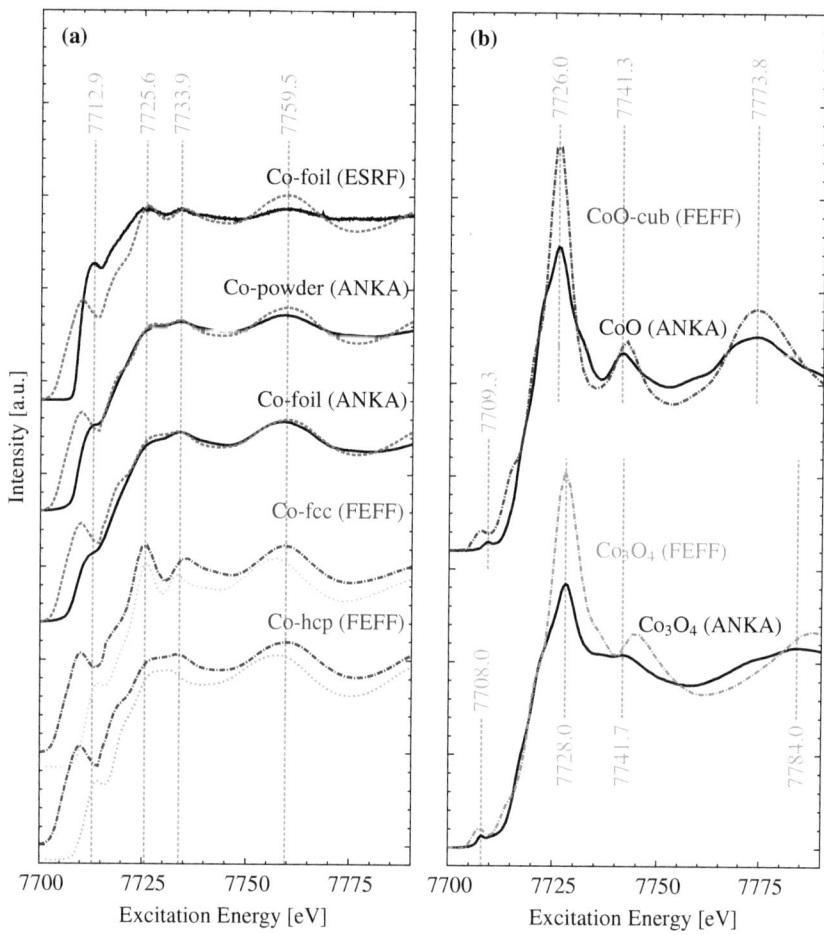

Figure C.2.: Normalized Co K-edge XANES simulations by FEFF9 in comparison to respective references. (a) Experimental Co metal data (black lines) with LCFs (red short-dashed lines) by Co-fcc and Co-hcp (with exchange potential DH+HL: dash-dotted lines). Additionally Co-fcc and Co-hcp calculated with DH potential is shown (dotted lines). (b) Experimental CoO and Co_3O_4 data (black lines) as well as CoO-cub and Co_3O_4 (-diamond). All significant features are marked by vertical dashed lines and the respective energies are given in eV.

D. EXAFS

Background removal

Before an EXAFS spectrum is visualized in position space, a background removal will be performed. For this a smooth background that reflects the "low frequency Fourier components" will be fitted to the normalized (explained in appendix C) XAS spectrum. The origin of this background function lies mainly in photoelectron scattering at the electronic environment of the absorbing atom, the so called atomic XAFS or AXAFS [36]. AXAFS, albeit bearing some useful information [73] which, however, is beyond the scope of this work, can not be simulated by the theory that is utilized for the EXAFS analysis. Furthermore, AXAFS disturbs the "real" EXAFS, which stems from photoelectron scattering at the atomic environment of the absorbing atom and will thus be removed. The background function is determined in R-space, by the software Athena [69], where a "cut-off" parameter R_{bckg} can be set. All R-space oscillations up to R_{bckg} are fitted and back Fourier transformed to E-space, as can be seen in Fig. D.1, where the subsequent steps of the background removal procedure are shown, too. Additionally the transformations to k- as well as R-space are also shown without background removal. For more details about this issue the reader is referred to the documentation of Athena software by B. Ravel [68].

EXAFS analysis

EXAFS spectra are commonly given in k-space and/or R-space. The k-space spectra $\chi(k)$ represent the oscillations originating from the atomic neighborhood only (by means of the "background-removal" explained previously), without the contributions from an isolated atom and its atomic X-ray absorption fine structure (AXAFS) [36], i.e. the bare edge-jump and the scattering of the photo-electron from the electronic environment of the atom. $\chi(k)$ is usually weighted by multiples of k to emphasize the low or high k-values, resulting in k-weighted $\chi(k)$, i.e. $k^n \cdot \chi(k)$ with typically $n = 1, 2$ or 3. $k^n \cdot \chi(k)$ then is Fourier-transformed to R-space (keeping the k-weight, see Eq. (D.1), wherefore a certain k range is chosen, with the aim to exclude bound states at low k values (k_{min}) and bad data quality when the oscillations become too weak at high k values (k_{max}).

$$\hat{\chi}(R) = \int_{k_{min}}^{k_{max}} dk \; k^n \, \chi(k) \, e^{-2\pi i k R} \tag{D.1}$$

This k-range is realized by a window-function. A typical example is given in Fig. D.2 (a) and (b) for Co-foil with a k-range from 3 to 13 Å^{-1}. As the amplitude in R-space is decreasing with increasing R, according to the EXAFS equation

$$\chi(k) = S_0^2 \sum_j \frac{N_j f_j(k)}{k R_j^2} e^{-2k^2 \sigma^2} e^{-2R_j/\lambda(k)} \, sin[2kR_j + \phi_j(k)], \tag{D.2}$$

D. EXAFS

the $\chi(R)$ spectrum is typically restricted to about 6 Å. Here $R_j = R_0 + \Delta R$, and $k^2 = 2m(E - E_0)/\hbar$ ($E_0 = E_0^{calculated} + \Delta E_0$), compare Eq. (2.33) on page 17 and explanations given there. This R-range can be regarded by a window-function, which can be further utilized to back-(Fourier-)-transform $\chi(R)$ to get the $q^n \cdot \chi(q)$ spectrum that reflects the EXAFS oscillations effectively used for the chosen R-range. Here it is important to realize that $\chi(R)$ actually consists of a real and imaginary part, as the oscillations in k-space exhibit positive and negative values. All this is demonstrated in Fig. D.2 (a) and (b) for Co-foil, where the R-range is chosen from 1 to 5 Å.

The EXAFS fits are performed with the help of the ARTEMIS software [69] that provides a graphical user interface for FEFF [75] with the focus onto EXAFS. The standard settings for the EXAFS simulations are given in D.1. Briefly, a crystal sphere of radius 8.0 Å is calculated around the absorbing Co atom according to Eq. (2.25) (section 2.2.2) and its k-space formulation Eq. (D.2), respectively, i.e. each order (up to number of "legs" NLEG = 4) of multiple scattering contributions - also named "paths", as it is the propagation and scattering path of the photo-electron - is calculated separately. However, the various contributions are not summed up immediately like in Eq. (2.25), instead one can visualize (in k- or R- space) and compare each path separately with the experimental data, wherefore the blue terms in Eq. (D.2) are calculated by FEFF and the green terms (parameters) are set to initial default values. In the fitting procedure, these EXAFS parameters are adjusted and the paths are finally summed up to reconstruct the experimental spectrum. Here it is important to know the number of independent points (IP) attributed to the experimental spectrum and to which the parameters are fitted. These IP are equivalent to the number of points, necessary to determine a spline for the experimental spectrum, according to:

$$IP = 2\Delta k \cdot \Delta R/\pi. \tag{D.3}$$

Here Δk is the k-range of the Fourier transform (3 to 13 reciprocal Å) and ΔR is the fitting range in R-space (1 to 5 Å).

All EXAFS fits are performed in R-space with respect to the real and imaginary part, not in k-space, since only a part of the k-space oscillations is utilized for the R-space spectra as was demonstrated in Fig. D.2. The EXAFS fits of the metallic Co references by simulations of Co-fcc or Co-hcp, are shown in Fig. D.3 and Fig. D.4 and in Table D.2. In Fig. D.3 the real and imaginary part of the R-space spectra are also given besides the magnitude to get the full picture. In the left column of Fig. D.3 the single-scattering (ss) paths are explicitly shown and in (b) the significant double- and triple-scattering (ds/ts) paths. The single scattering paths 1, 2, 5 and 8 are reflecting the Co shells that are surrounding the central absorbing Co atom and are clearly visible in $|\chi(R)|$ of Co-foil. The amplitude of the strongest 1st shell path is set to 100 by FEFF and all paths with amplitude ≥ 10 were included in the EXAFS fits performed in this work.

Table D.1.: Standard FEFF input file for EXAFS simulations.

EDGE	K					
S02	1.0					
COREHOLE	rpa					
*	pot	xsph	fms	paths	genfmt	ff2chi
CONTROL	1	1	1	1	1	1
PRINT	1	0	0	0	0	3
*	ixc	vr0	vi0	[ixc0]		
EXCHANGE	0	0	0	0		
*	kmax					
EXAFS	20.0					
RPATH	8.0					
NLEG	4					
*	emin	emax	eimag			
LDOS	-30	20	0.1			
POTENTIALS						
*	ipot	Z	[tag]	[lmax1]	[lmax2]	[xnatph]
	0	27	Co	2	2	0.001
	1	27	Co	2	2	1
ATOMS						
	(list of xyz coordinates of atom-cluster)					

D. EXAFS

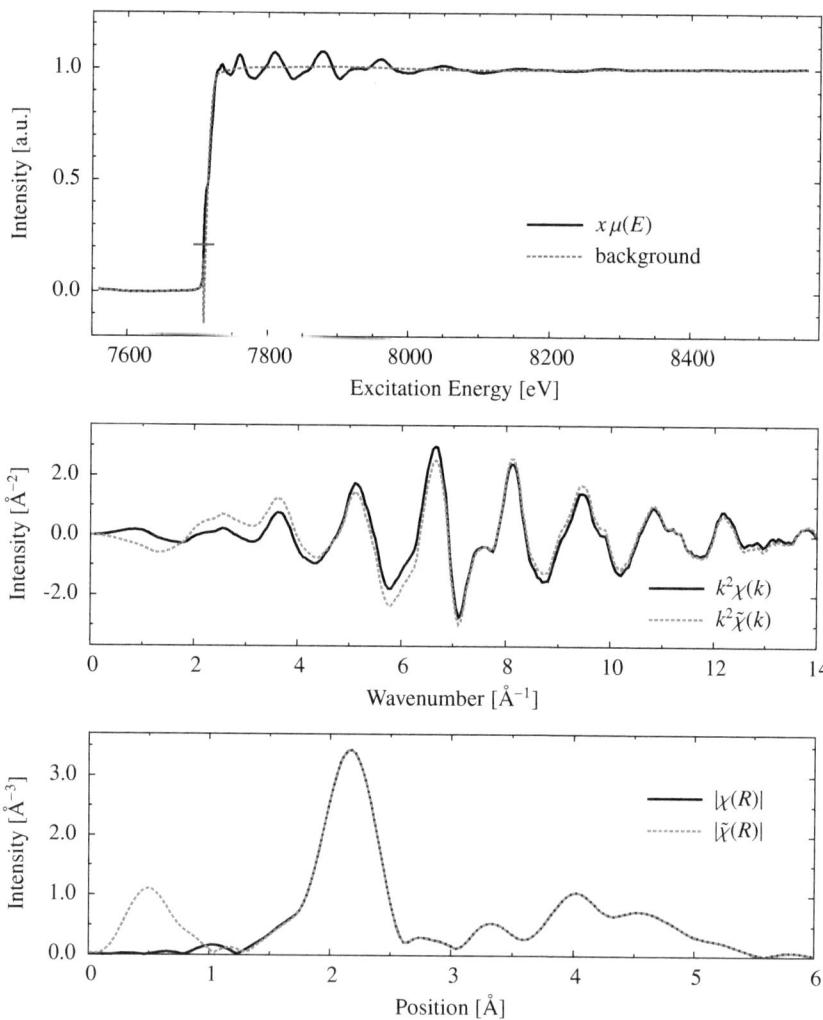

Figure D.1.: Top: Normalized XAS spectrum $x\mu(E)$ of a Co powder measured in transmission along with background function. Middle: $x\mu(E)$ after background removal and transformation to k-space and with k^2-weight, $\chi(k)$, and without background removal, $\tilde{\chi}(k)$. Bottom: Fourier transformation $\chi(k)$ and $\tilde{\chi}(k)$ to R-space.

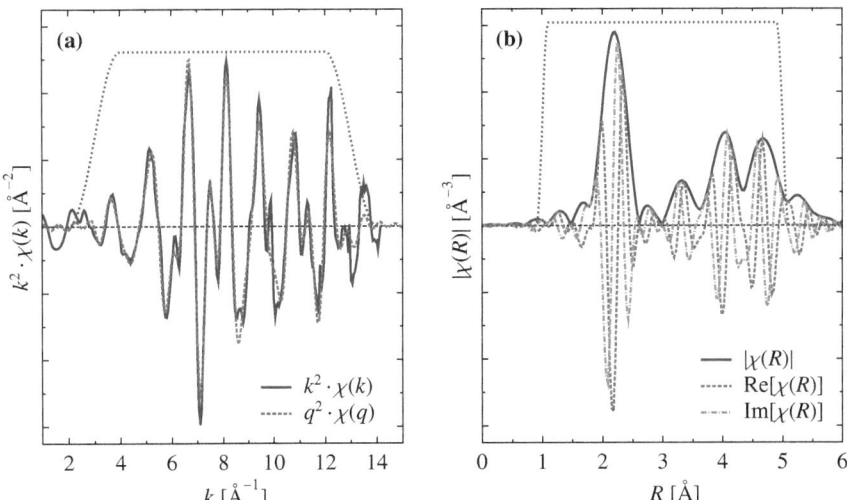

Figure D.2.: HRFD-EXAFS spectrum of Co-foil. (a) k^2-weighted k-space spectrum $\chi(k)$ along q^2-weighted with the q-space spectrum $\chi(q)$. The latter is the back transformation of the R-space spectrum in (b) within the window given there. (b) R-space spectrum, Fourier transformed from $k^2 \cdot \chi(k)$ in (a) within the window given there. Magnitude $|\chi(R)|$, real part $\text{Re}[\chi(R)]$ and imaginary part $\text{Im}[\chi(R)]$ are shown.

D. EXAFS

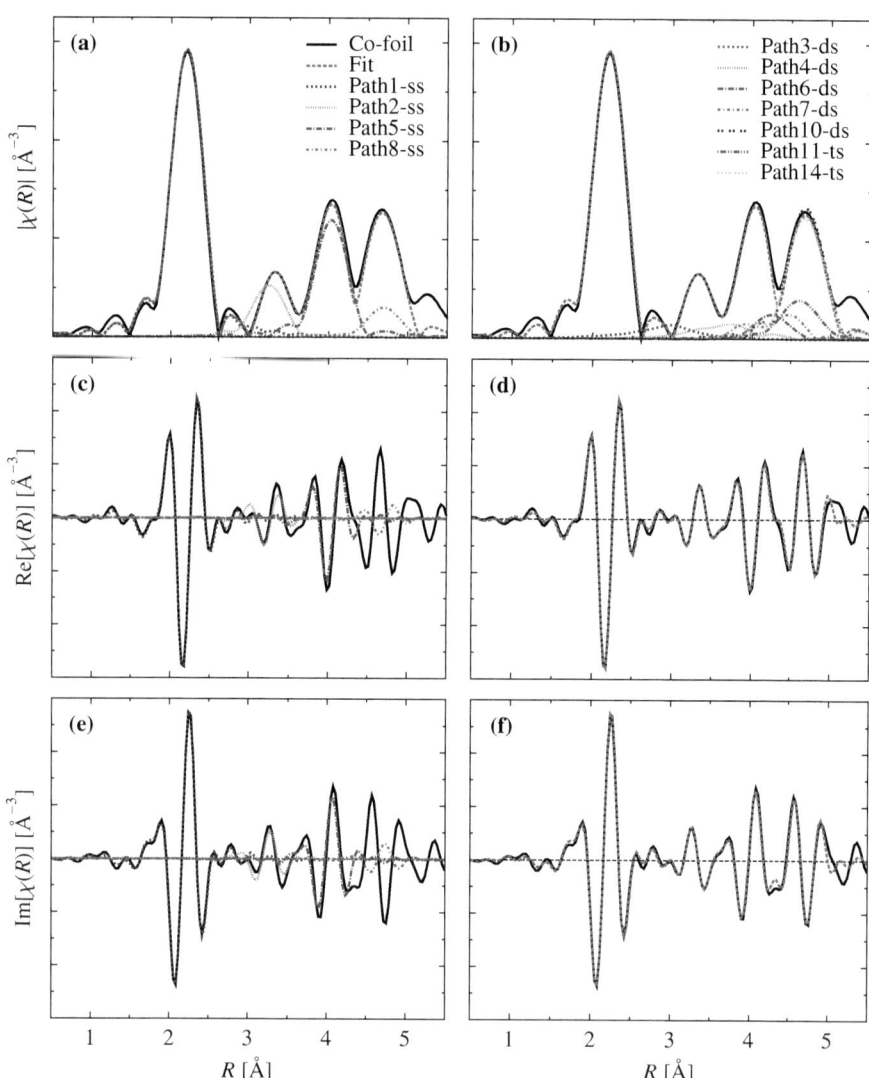

Figure D.3.: Fit of R-space HRFD-EXAFS spectrum of Co-foil-ESRF by FEFF simulation of Co-fcc. Magnitude $|\chi(R)|$ of Co-foil, Fit and single-scattering (ss) paths (a) and double-/triple-scattering (ds/ts) paths (b). Real and imaginary part of $\chi(R)$ of Co-foil and ss paths in (c) and (e) and respective fitting curves in (d) and (e).

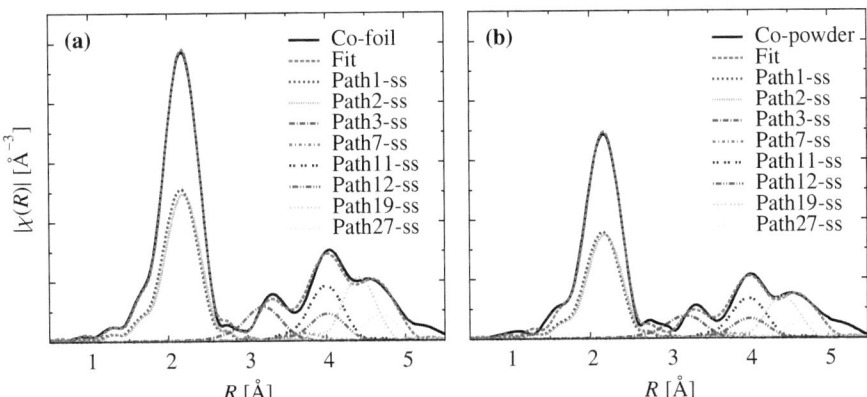

Figure D.4.: Fit of R-space transmission EXAFS spectra of Co-foil-ANKA (a) and Co-powder-ANKA (b) by FEFF simulations of Co-hcp. Magnitudes $|\chi(R)|$ of experimental data, Fit and single-scattering (ss) paths are shown.

D. EXAFS

Table D.2.: EXAFS fits of the metallic Co references used in this work. Amplitude reduction factor S_0^2 is determined initially for the 1st shell of Co-foil to 0.76 and then set constant for all fits. The additional factor δN reflects the amplitude reduction due to differences in the measurements.

experimental data		Co-foil (ESRF)	Co-foil (ANKA)	Co-powder (ANKA)
Feff simulation		Co-fcc	Co-hcp	Co-hcp
R-factor [$\times 10^{-3}$]		8	2	6
Energy shift	E_0 [eV]	11.0(4)	9.7(3)	9.7(4)
Reduction factor	δN	1.00(4)	1.00(3)	0.58(3)
Debye-Waller-factors σ^2 [10^{-3}Å^2]	shell-1	2.5(03)	6.0(02)	5.7(04)
	shell-2	3.5(09)	10.9(10)	11.1(16)
	shell-2/3	–	10.9(10)	11.1(16)
	shell-3	3.6(05)	7.8(04)	7.3(07)
	shell-3/4	–	4.8(06)	4.9(11)
	shell-4	5.4(12)	5.1(17)	5.7(26)
shell distances d [Å]	shell-1	2.499(2)	2.484(3)	2.484(05)
	shell-2	3.534(3)	3.513(4)	3.512(07)
	shell-2/3	–	4.011(7)	4.018(11)
	shell-3	4.328(4)	4.318(3)	4.315(06)
	shell-3/4	–	4.725(6)	4.730(10)
	shell-4	4.998(4)	4.995(4)	4.990(26)
lattice constants	absolute values [Å]	3.534(3)	2.497(2)	2.495(03)
			4.011(7)	4.018(11)
	relative deviation [%]	-0.3(1)	-0.4(1)	-0.5(1)
			-1.4(2)	-1.3(3)

List of Figures

2.1. Simple scheme of photon interaction with matter. Here x is the materials thickness and I_0 and I is the initial and transmitted intensity, respectively. 6
2.2. Linear mass attenuation coefficient μ/ρ for cobalt on a double-logarithmic scale. . . . 7
2.3. The photoelectric effect where a X-ray photon is absorbed by a core-shell electron that is ejected, i.e. excited into the continuum (left). This is followed by the relaxation of the atom via X-ray fluorescence (K_α or K_β) (middle) or the Auger effect (right) (picture taken from [60].). 9
2.4. Schematic K-edge X-ray absorption spectrum of a transition metal compound and its division into three regions. Dominant electron transitions (1s to $n = 3, 4, 5$) are assigned to each region (taken from [100]). 10
2.5. The atom is hit by a X-ray (left). Hereby, a photoelectron is ejected in form of a spherical wave (middle) and interferes with its scattering part (right). 11
2.6. XAS through the photoelectric effect for a single atom (blue lines only) and for an atom surrounded by other atoms (blue and red lines): The ejected photoelectron is scattered from a neighboring atom. The scattering electron wave, when returning to the absorbing atom, interferes with the initial photoelectron wave, which modulates the absorption probability (picture taken from [60]). 12
2.7. Muffin-tin approximation: Non-overlapping spherical potentials separated by interstitial regions of constant potential (picture taken from [55]). 13
2.8. Visualization of single-, double-, and triple-scattering paths for three fixed atoms, as described by the full Green's function Eq. (2.25). 15
2.9. 2nd order optical process as described by the Kramers-Heisenberg equation Eq. (2.34) for the case of cobalt. Co has the ground state configuration [Ar]$3d^74s^2$ ($4s^2$ electrons not drawn) or short: $|g\rangle = 3d^7$ + X-ray. The intermediate state is $|n\rangle = 1s3d^8$ (1s stands for the core-hole) and the final state is $|f\rangle = 3p^53d^8$ + X-ray. 18
2.10. Schematized XES transitions for Co that are appearing in this work. On the left is given the electron binding energy E_b (taken from the NIST database [61]) and on the right the edge label and in brackets the corresponding electron orbital. 19
2.11. 1s3p RIXS plane of CoO (measured at ID26, ESRF, see section 6.3) (top-right). High-resolution Co $K\beta_{1,3}$ RXES spectrum extracted at excitation energy of 7726.0 eV on the left (vertical white dashed line in RIXS plane) and Co K-edge HRFD-XANES spectrum extracted at emission energy of 7750.8 eV at the bottom (horizontal white dashed line in RIXS plane). 20
2.12. 1s-3p RIXS plane of CoO. Also shown are the orientations of the core-hole lifetime broadenings Γ_{1s} and Γ_{3p} as well as the direction for extraction of a HRFD-XAS spectrum, labelled $I(\omega_{in}, \omega_{out}=7651\,eV)$. Top: Emission against excitation energy is shown. Bottom: Energy transfer against excitation energy is shown. 21
2.13. Normalized $K\beta_{1,3}$ emission line of Co-foil and CoO (as powder). ΔE denotes the chemical shift due to different valencies of the two shown samples. 22

List of Figures

2.14. Left: Section of the 1s3p RIXS plane of Co-foil. Emission against excitation energy is shown. The arrows indicate the direction where the HRFD-XANES spectra are extracted from. Right-top: The formerly described (normalized) HRFD-XANES spectra. Right-bottom: Co-foil HRFD-XANES spectrum with standard deviation, averaged from the three Co-foil spectra shown in the upper figure. 23

3.1. Typical scheme of a synchrotron radiation facility (taken from [107]), with its different stations labelled and explained in the text. 26

3.2. Schematic description of a wiggler (a) and an undulator (b) as well as their spectral brilliance (in photons/s/mm^2/mrad^2/0.1%BW, where 0.1%BW denotes a bandwidth $10^{-3}v$ centered around the frequency $v = c/\lambda$, with c the speed of light and λ the photons wavelength) in comparison to a normal bending magnet (c) (figures taken from [35]). 27

3.3. "Bragg diffraction. Two beams with identical wavelength and phase approach a crystalline solid and are scattered off two different atoms within it. The lower beam traverses an extra length of $2 \cdot d \sin\theta$. Constructive interference occurs when this length is equal to an integer multiple of the wavelength of the radiation." [99] 28

3.4. Schematized setup of the INE beamline at ANKA (a), with photographs of the experimental stage (b) as well as of the double-crystal X-ray monochromator (c). 29

3.5. Schematized setup of HASYLAB's W1 beamline at DORIS III (taken from [33]). . . 30

3.6. Schematized Johann spectrometer: Detection of fluorescent X-ray radiation in Rowland geometry (circle of radius R) by a spherically bent crystal analyzer with radius of curvature $2R$. 30

3.7. Rowland geometry as utilized at the ID26 beamline at the ESRF. The red lines denote the fluorescent radiation which is emitted from the sample and then Bragg diffracted by five analyzer crystals and focused onto the detector (taken from [28]). 31

4.1. Two general approaches in nanoparticle synthesis. 33
4.2. Possible cobalt phases: Co-hcp, Co-fcc, Co-ε (β-Manganese), and Co-bcc. 34
4.3. TEM picture of the Co nanoparticles. 36
4.4. HRTEM images of Co$_3$Pt/C nanoparticles in the states 'as prepared' (a) and tempered to 800 °C (b) (pictures taken from [81]). 37

5.1. Diagram of a proton exchange membrane fuel cell (PEMFC) (picture taken from [105]). 40
5.2. Normalized Co K-edge XANES spectra of references Co-foil and CoO (a) (edge shoulder and whiteline marked by arrows), as well as CoO, Co$_3$O$_4$ and Co$_2$O$_3$ (b) (with whiteline positions marked by vertical lines). 41
5.3. Normalized Co K-edge XANES spectra of the Co$_3$Pt/C nanoparticle catalyst 'as prepared' as well as tempered to 350 °C and 800 °C in comparison to references. 42
5.4. (a) Normalized Pt L3-edge XANES spectra of the Co$_3$Pt/C nanoparticle catalyst 'as prepared' as well as tempered to 350 °C and 800 °C (upper curves) and the latter compared to commercial Pt$_3$Co/C catalyst from TKK (lower curves). (b) Normalized Pt L3-edge XANES spectra of references Pt-foil and PtO$_2$. 43
5.5. EXAFS spectra of the Co$_3$Pt/C nanoparticle catalyst in three states at both edges. k^2-weighted k-space spectra on the left and respective Fourier transformed ($FT \longrightarrow$) R-space spectra on the right. 45

List of Figures

5.6. Simultaneous fit of Pt L3- and Co K-edge EXAFS spectra of Co_3Pt/C-800 by FEFF [75] simulated PtCo-fcc, with main fit contributions (scattering paths) for each coordination shell (C, 1, 2, 3, 4 and 5) included. 46

5.7. Lattice constant a_i for Co_3Pt/C at 800 °C for coordination-shells i, as derived from Pt–Pt, Pt–Co, and theoretical path distances of the simultaneous EXAFS fit. 47

5.8. Fit of Pt L3-edge EXAFS spectra of Co_3Pt/C-800 and Pt_3Co/C-TKK by a disordered PtCo-fcc phase. The labels C and 1 to 5 denote the coordination-shells, whereof the contributions to the first shell are shown. 48

5.9. Diffusion-corrected specific ($i_{kin,s}$) as well as mass activity ($i_{kin,m}$) for the ORR as determined by RDE measurements at 0.9 V [81] (a) in comparison to selected EXAFS fit quantities for Co_3Pt/C-asprep, 350, 800 and Pt_3Co/C-TKK (b - d). 49

6.1. Normalized $K\beta_{1,3}$ emission spectra of Co, CoO, Co_2O_3 and the Mix (78.1 % Co and 21.9 % CoO), all measured as powders. The insets show a magnification to both peaks $K\beta'$ and $K\beta_{1,3}$. 57

6.2. Normalized and aligned $K\beta_{1,3}$ emission spectra (blue filled circles) of powdered Co (left) and CoO (right) fitted by three Voigt functions (black dash-dotted curves and total fit as red dashed curve) which do not represent real resonances. Fit details are given in Table 6.1. 58

6.3. Top: Normalized $K\beta_{1,3}$ NRXES spectra of powdered samples Mix, Co and CoO. The latter two spectra's intensities are weighted with respect to their contribution in Mix, as determined by the LCF. Bottom: Fractions of Co and CoO in Mix, as calculated by the LCF. The arrows (top and bottom figure) denote the mean energies of the intervals with intermediate (Pos-1), highest (Pos-2), and lowest (Pos-3) Co to CoO ratio, as chosen for the extraction of the HRFD-XANES spectra. 60

6.4. Normalized Co K-edge HRFD-XANES spectra. Left: The Mix, extracted from fluorescence regions denoted in brackets (and indicated by arrows in Fig. 6.3). The two arrows indicate the regions that are most sensitive to the Co valency, i.e. Co^0 and Co^{2+} in this case. Right: Comparison of the Mix spectrum with those of its components Co and CoO, all extracted from the interval at Pos-1. 61

6.5. LCF of the three HRFD-XANES spectra of the Mix. For each fit the data (black line) is shown along with the fit (red slashed) and the fit components Co (blue dotted) and CoO (light-blue dash-dotted). The two fit components are taken from the same intervals as Mix for each fit. 62

6.6. Normalized HRFD-XANES spectra of Co and CoO extracted at the $K\beta_{1,3}$ emission peak (peak), in comparison to the average of the HRFD-XANES spectra extracted from the three intervals (avg). 63

6.7. Normalized Co K-edge spectra of Co and CoO as obtained from the SVD in comparison to their experimental HRFD-XANES spectra Co and CoO ("avg" denotes the average of the three HRFD-XANES spectra). 64

6.8. Simple spherical core-shell model. A metallic Co core surrounded by a Co-O/C shell of valency 2 and in between a transition layer. 65

6.9. Top: $K\beta_{1,3}$ emission lines of Co-nano, Co-nano-ox and Co-foil. The latter two spectra are scaled with respect to their contribution in Co-nano, as calculated by the LCF. Bottom: Fractions of Co-nano-ox and Co-foil in Co-nano, as calculated by the LCF. Intervals (shown as dashed boxes) denote regions with intermediate, lowest and highest shell to core ratio. 66

List of Figures

6.10. Normalized Co K-edge HRFD-XANES spectra. Left: Co-nano, extracted from fluorescence regions (shown in 6.9 bottom) with most distinct shell rates as denoted in brackets. The two arrows indicate the regions that are sensitive to the Co valency. Right: Co-nano and its model compounds extracted at the "high" interval. 67

6.11. Normalized HRFD-XANES spectra of the model compounds extracted at the $K\beta_{1,3}$ emission peak (peak), in comparison to the average of the HRFD-XANES spectra extracted from the three intervals (avg). 68

6.12. Final set of site-selective XANES spectra for the two sites, core and shell, of the Co nanoparticles, as obtained from the numerical procedure and in consideration of the restraints. Five equidistant steps in between the final parameter range are applied for both sites. 69

6.13. Normalized Co K-edge XANES spectra of the Co nanoparticles' core (with standard deviation) in comparison to the averaged Co-foil (left) and in comparison to FEFF 8.4 simulated ε, hcp and fcc structures of Co (with intensity offset). 70

6.14. Normalized Co K-edge XANES spectra. Left: The shell (with standard deviation) of the Co nanoparticle in comparison to the averaged Co-nano-ox. Right: The shell of the Co nanoparticle as well as a cobalt(II)-oxide (CoO) and a cobalt(II,III)-oxide (Co_3O_4). The latter two spectra are offset with respect to the intensity. The vertical line at 7708.5 eV indicates the position of the pre-edge feature of the average shell spectrum. The insets show a magnification of the pre-edge regions. 71

6.15. Normalized Kβ emission spectra of the three references Co, CoO and $CoCO_3$. The main $K\beta_{1,3}$ line and Kβ′, as well as the ligand sensitive $K\beta_{2,5}$ line and crossover resonance Kβ″ are shown. The inset shows a zoom to the latter, with ×35 magnification. 74

6.16. Normalized Co K-edge HRFD-XANES spectra of (left) Co, CoO and $CoCO_3$, extracted at their $K\beta_{1,3}$ peak positions (see point 3. in Table 6.5), and (right) the same CoO spectrum in comparison to its total fluorescence and transmission scan. 75

6.17. Normalized NRXES $K\beta_{1,3}$ spectra of the three references (left) and the three nanoparticles (right). The two arrows indicate the peak position of Co-foil and CoO, respectively. For the nanoparticles' spectra both peaks are also shown magnified. 76

6.18. First three panels: LCF's of the three $K\beta_{1,3}$ NRXES spectra of the Co nanoparticles by the references Co and CoO. The bottom panel shows the fractions of Co (first three curves) and CoO (last three curves) in Co-nano-1, 2 and 3 as determined by the LCF's. The three arrows and the vertical dashed lines, respectively, indicate the positions from which the HRFD-XAS spectra are extracted. 77

6.19. Normalized Co K-edge HRFD-XANES spectra of the Co nanoparticles. Left panel (Method 1): Comparison of spectra from Pos-2, 1 and 3 of Co-nano-1 (a), 2 (b) and 3 (c) respectively. Right panel (Method 2): Comparison of spectra of Co-nano-1, 2 and 3 measured at Pos-1 (d), 2 (e) and 3 (f). The spectra with labels written in cursive are excluded in the further process as explained in the text. 78

6.20. Normalized HRFD-XANES spectra of the three references Co, CoO and $CoCO_3$ in comparison to their respective averaged spectra from Pos-2 and 3 (and also from Pos-1, 2 and 3 for Co-foil). 80

6.21. Site-specific Co K-edge spectra of the Co nanoparticles core (left) and shell (right) respectively. The spectra are calculated from HRFD-XANES spectra as given in brackets in the labels and corresponding to the methods 1, 2 and S. 81

List of Figures

6.22. Left (Right): Site-specific XANES spectrum of the Co nanoparticles' core (shell) as calculated by SVD (method given in brackets) along with its model compound Co-foil (CoO) at the bottom and along with the HRFD-XANES spectra of the Co nanoparticles from Pos-2 (Pos-3) at the top. 83

6.23. The site-specific XANES spectrum of the Co nanoparticles' core in comparison to normalized Co K-edge XANES spectra of the metal Co phases, simulated by FEFF9 (a) and FDMNES (b) and the Co-foil. The vertical dashed lines mark the significant features of the core spectrum. A magnification of a part of the core spectrum along with Co-hcp is given at the bottom. 85

6.24. The site-specific XANES spectrum of the Co nanoparticles' shell in comparison to normalized Co K-edge XANES spectra of Co-O/C references, measured at the same beamline (a) and elsewhere in transmission (b). Significant features of the shell spectrum are marked by dashed vertical lines. 86

6.25. Normalized Co K-edge XANES spectra of CoO and Co_3O_4 (both measured in transmission) along with respective simulations calculated by FEFF (a) and FDMNES (b). All pre-edge regions are also shown magnified whereby the position of the (center of) the peak is preserved, but its shape is stretched about a factor 2. Significant features of the experimental spectra are marked by dashed vertical lines. 87

6.26. Site-specific Co K-edge XANES spectra of the Co nanoparticles as obtained by SVD with 3 components (a) and with 4 components (b). The Co references are also plotted. The vertical dashed line marks the pre-edge position of CoO (avg) at 7709.3 eV. . . . 90

6.27. Left: k^2-weighted k-space EXAFS spectra of the references Co-foil, CoO and $CoCO_3$. Right: Phase-corrected magnitudes of the R-space EXAFS spectra, fourier transformed from k-space spectra on the left. The labels are denoting the spectra on the left likewise. 92

6.28. Left: k^2-weighted k-space HRFD-EXAFS spectra of the nanoparticles Co-nano-1, 2 and 3. Right: Magnitudes of the R-space HRFD-EXAFS spectra, fourier transformed from k-space spectra on the left. The labels are denoting the extraction position for the spectra (on the left and right likewise). 93

6.29. Site-specific EXAFS spectra for the core and shell as determined by the SVD. The labels are denoting the methods, as explained in the text, for the left and right spectra likewise. The k^2-weighted k-space spectra on the left are fourier transformed to give the (magnitudes of the) R-space spectra on the right. 95

6.30. Left (Right): Site-specific EXAFS spectra for the core (shell) in comparison to the nanoparticles' HRFD-EXAFS spectra averaged at Pos-2 (Pos-3). The core (shell) spectrum of method 1 is also averaged. Shown is the magnitude of $\chi(R)$ on top and its real part at the bottom. 96

6.31. Site-specific EXAFS spectrum of the core of method S in comparison to the spectra of the FEFF simulated metal Co-phases (no fit, but downscaled by factor 0.15) in k-space (left) with k^2-weight and in R-space (right) fourier transformed from spectra on the left. 97

6.32. Core spectrum from method S fitted by Feff simulation of Co-hcp (left) and Co-fcc (right). Also shown are the single-scattering (ss) paths, which reflect the coordination-shells, marked by boldface numbers. 98

143

List of Figures

6.33. HRFD-EXAFS spectrum (left) and site-specific shell spectrum (right) from method S fitted by Feff simulation of rocksalt CoO and Co-hcp. Also shown are the single-scattering (ss) paths, which reflect the coordination-shells, marked by boldface labels/numbers. 100

6.34. VTC-XES spectrum of CoO with labels for the peaks on the left and which origins are schematized at the right ("L" denotes the ligand). 102

6.35. FEFF l-DOS simulations in comparison to experimental VTC-XES spectra (black data points •): (a) rocksalt CoO-cub as well as (b) Co_3O_4 along with CoO on the left and (c) Co_2C as well as (d) $CoCO_3$ along with $CoCO_3$ on the right. 104

6.36. $K\beta_{2,5}$-RXES (or VTC-XES) spectra of the references Co, CoO and $CoCO_3$ (left) as well as of the three Co nanoparticles (right). The arrows indicate the significant features of the references, with labels denoting their origin. 105

6.37. $K\beta_{2,5}$-XES spectra of Co-nano-1 and Co-nano-3 fitted by the references Co-foil, CoO and $CoCO_3$. The reference spectra are scaled with respect to the fitting results. The vertical dashed lines denote the positions of the significant features of CoO and $CoCO_3$ in the low-energy range, as identified by FEFF simulations (see Fig. 6.35). . 106

6.38. FEFF $K\beta_{2,5}$-XES simulations of (a) Co metals, (b) Co-oxides, (c) Co-nitrogens and (d) Co-carbons, in comparison to Co-nano-3 (gray data points •) and its LCF (red dashed line - - -) and one of the references (a) Co-foil, (b) CoO and (d) $CoCO_3$ (black data points •). The vertical lines mark significant features of the references and the simulations, respectively. 107

6.39. $K\beta_{2,5}$-XES spectra of the three references (left) and the three nanoparticles (right). The spectra are cut off from the normalized $K\beta_{1,3}$ NRXES spectra, whose high-energy tail is visible therefore. 109

6.40. Normalized HRFD-XANES spectra of the three nanoparticles from fluorescence, detected at the $K\beta_{2,5}$ peak positions of CoO in the left panel (Pos-4) and of Co in the right panel (Pos-5). The spectrum of Co-nano-1 is compared to that of Co-foil and the site-specific core (see Fig. 6.22) for Pos-4, and the region from 7722 to 7774 eV is magnified with respect to its intensity. The Co-Nano-3 spectrum from Pos-5 is compared with Co-foil and CoO. 110

A.1. Ratio of surface atoms to bulk atoms according to Eq. (A.5) for Pt and Co on a linear scale. 119

A.2. Ratio of surface atoms to bulk atoms for Palladium, taken from [63] with logarithmic scale on the x-axis. 120

C.1. Raw experimental XAS spectrum for a Co powder measured in transmission (top), along with pre-edge line and post-edge polynomial that are used to determine the normalized spectrum at the bottom. 125

C.2. Normalized Co K-edge XANES simulations by FEFF9 in comparison to respective references. (a) Experimental Co metal data (black lines) with LCFs (red short-dashed lines) by Co-fcc and Co-hcp (with exchange potential DH+HL: dash-dotted lines). Additionally Co-fcc and Co-hcp calculated with DH potential is shown (dotted lines). (b) Experimental CoO and Co_3O_4 data (black lines) as well as CoO-cub and Co_3O_4 (-diamond). All significant features are marked by vertical dashed lines and the respective energies are given in eV. 130

List of Figures

D.1. Top: Normalized XAS spectrum $x\mu(E)$ of a Co powder measured in transmission along with background function. Middle: $x\mu(E)$ after background removal and transformation to k-space and with k^2-weight, $\chi(k)$, and without background removal, $\chi(\tilde{k})$. Bottom: Fourier transformation $\chi(k)$ and $\chi(\tilde{k})$ to R-space. 134

D.2. HRFD-EXAFS spectrum of Co-foil. (a) k^2-weighted k-space spectrum $\chi(k)$ along q^2-weighted with the q-space spectrum $\chi(q)$. The latter is the back transformation of the R-space spectrum in (b) within the window given there. (b) R-space spectrum, Fourier transformed from $k^2 \cdot \chi(k)$ in (a) within the window given there. Magnitude $|\chi(R)|$, real part $\text{Re}[\chi(R)]$ and imaginary part $\text{Im}[\chi(R)]$ are shown. 135

D.3. Fit of R-space HRFD-EXAFS spectrum of Co-foil-ESRF by FEFF simulation of Co-fcc. Magnitude $|\chi(R)|$ of Co-foil, Fit and single-scattering (ss) paths (a) and double-/triple-scattering (ds/ts) paths (b). Real and imaginary part of $\chi(R)$ of Co-foil and ss paths in (c) and (e) and respective fitting curves in (d) and (e). 136

D.4. Fit of R-space transmission EXAFS spectra of Co-foil-ANKA (a) and Co-powder-ANKA (b) by FEFF simulations of Co-hcp. Magnitudes $|\chi(R)|$ of experimental data, Fit and single-scattering (ss) paths are shown. 137

List of Tables

2.1. Possible electron states for the case of the principal quantum number $n = 2$. l is the orbital angular momentum quantum number, j the total angular momentum, $2j+1$ the multiplicity and nl_j the term symbol. In the last column the completely filled states are shown with electrons symbolized by arrows (↑ = spin up, ↓ = spin down). 8

3.1. Main parameters of the three synchrotrons. 27
3.2. Main parameters of the visited beamlines hosted at the synchrotrons and described in Table 3.1. For more information about the crystals used, see Table 3.3. 32
3.3. Crystals used by DCMs and spectrometers in this work. a is the lattice constant and $d = a/\sqrt{h^2 + k^2 + l^2}$ the interplanar distance with respect to the cut defined by the miller indices h, k, l. The energy ranges possible with these crystals are calculated based on Eq. (3.2), with $n = 1$ and for the DCM with $\theta = 15° - 65°$ (E_{ex}) and for the Johann spectrometer with $\theta = 60° - 86°$ (E_{em}). 32

5.1. EXAFS fits of Co$_3$Pt/C nanocatalysts and references at both edges. Part 1. 51
5.2. EXAFS fits of Co$_3$Pt/C nanocatalysts and references at both edges. Part 2. 52
5.3. EXAFS fits of Co$_3$Pt/C nanocatalysts as well as of Pt$_3$Co/C-TKK commercial catalyst at the Pt L3-edge, in comparison to ORR activities at 0.90 V as determined by RDE measurements [81]. 53

6.1. Results for the 3-Peak-Voigt-Fit of the four K$\beta_{1,3}$ emission spectra. Each column lists the energetic positions of the three Voigt-functions. The corresponding position at the K$\beta_{1,3}$ line is denoted in brackets. K$\beta_{1,3}^{\text{eff}}$ is the peak of the whole Voigt-fit, i.e. of the sum of the three Voigt-functions and ΔPeaks is the energetic difference between Peak-eff and Peak-1, i.e. the K$\beta_{1,3}$ to Kβ' splitting. 59
6.2. Details about K$\beta_{1,3}$ fluorescence intervals chosen for the extraction of HRFD-XANES spectra. ΔE is the width of the chosen intervals. 59
6.3. Results of all linear combination fits (LCF) of the Mix by Co and CoO. K$\beta_{1,3}$-LCF ratios from Table 6.2 with standard deviation given in brackets. XANES-LCF: Each Mix spectrum was fitted by Co and CoO extracted from the same interval. XANES-LCF2: Each Mix spectrum was fitted by the same Co and CoO spectra which are merges from the three intervals. Last column shows ratios as fitted by SVD. For the latter three fits the fitting errors are given in brackets. 62
6.4. Ratios Co-nano-ox : Co-foil in Co-nano, as obtained by LCF of the K$\beta_{1,3}$ line (standard deviation and fitting errors ±1) and as obtained by LCF of HRFD-XANES spectra (fitting errors ±0.5). Last two columns: Ratio shell : core in Co-nano as calculated by SVD: first the minimum and maximum ratios (with respect to the shell fraction), then the average ratios (errors negligible small about ±0.1). 67
6.5. Details about the various measurements at beamline ID26 described in section 6.3.1. 73

List of Tables

6.6. Overall Co-foil to CoO ratios as obtained via the LCF of the $K\beta_{1,3}$ spectra of Co-nano-1, 2 and 3, as well as for the specific positions Pos-1, 2 and 3. Below are the corresponding ratios from the LCF of the HRFD-XANES spectra. The R-factor describes the fit quality. The fit errors (given in brackets) of Pos-1, 2 and 3 are similar to those from the overall fits, correlated to the R-factor though. 75

6.7. Ratios of core : shell as obtained by the SVD of the Co nanoparticles' HRFD-XANES spectra from Pos-2 and 3. Results of three different methods (S, 1 and 2) are shown. The triplet of R-factors gives the fit quality for Co-nano-1, 2 and 3. The corresponding fit errors are always ±0.2 except for fits with R > 30 where it is ±0.4. 81

6.8. Ratios as obtained by LCF of the Co nanoparticles' HRFD-XANES spectra similar to Table 6.6, but with three fitting components Co, CoO and $CoCO_3$. The R-factors are describing the fit quality and the fit errors range between ±0.02 (R = 9) and ±1.0 (R > 40). SVD of the nanoparticles with spectra from Pos-2 and Pos-3 only: Listed are the ratios of the SVD with three (SVD-3) and with four (SVD-4) site-specific components core : shell-1 : shell-2 (: shell-3). 89

6.9. LCF of the Co nanoparticles' HRFD-EXAFS k-space spectra (k-weight = 0) in the k range 3 – 12 Å^{-1} with three fitting components Co, CoO and $CoCO_3$. The ratio errors are between ±2 and ±4, depending on the R-factors which are given for each three fits in a column. Below are the core : shell ratios as determined by two different SVD's. 94

6.10. EXAFS fits of HRFD-EXAFS spectra of Co-foil, of Co-nano-1+2+3 from Pos-2 simultaneously, of the site-specific core-1+2+3 from method 1 simultaneously and of the site-specific core from method S by FEFF simulations of Co-hcp (top) and Co-fcc (bottom). 99

6.11. EXAFS fits of the HRFD-EXAFS spectra of Co-nano-1, 2, 3 from Pos-3 simultaneously and of the site-specific shell from method S by FEFF simulations of Co-hcp and Co-O. 101

6.12. Results of the LCF of the three Co nanoparticles' $K\beta_{2,5}$-XES spectra by Co-foil, CoO and $CoCO_3$. The fit quality is given by the R-factor, and the fitting errors are given in brackets (with respect to last digit). 105

6.13. Results for the LCF of the nanoparticles' HRFD-XANES spectra from $K\beta_{2,5}$. The fitting components have been the site-specific core and shell spectrum as determined by method S in section 6.3.4. 111

C.1. Standard FEFF9 input file as utilized for all XANES simulations in this work. 127
C.2. LCFs of the three metallic Co K-edge XANES spectra by FEFF9 simulations of Co-fcc and Co-hcp (computational details in Table C.1). 128
C.3. Crystallographic data of all simulated Co compounds. 129

D.1. Standard FEFF input file for EXAFS simulations. 133
D.2. EXAFS fits of the metallic Co references used in this work. Amplitude reduction factor S_0^2 is determined initially for the 1st shell of Co-foil to 0.76 and then set constant for all fits. The additional factor δN reflects the amplitude reduction due to differences in the measurements. 138

Bibliography

[1] Fuel Cells 2000. Fuel Cell Basics: Applications.

[2] S. Alayoglu, A. U. Nilekar, M. Mavrikakis, and B. Eichhorn. Ru–Pt core–shell nanoparticles for preferential oxidation of carbon monoxide in hydrogen. *Nature Materials*, **7**:333–338, 2008.

[3] A. L. Ankudinov, B. Ravel, J. J. Rehr, and S. D. Conradson. Real-space multiple-scattering calculation and interpretation of x-ray-absorption near-edge structure. *Phys. Rev. B*, **58**(12):7565–7576, 1998.

[4] C. J. Ballhausen and Harry B. Gray. *Molecular Orbital Theory*. W. A. Benjamin, Inc., New York, 1956.

[5] J. L. Beeby. The Density of Electrons in a Perfect or Imperfect Lattice. *Proc. R. Soc. Lond. A*, **302**:113–136, 1967.

[6] P. Behrens. X-ray absorption spectroscopy in chemistry II: X-ray absorption near edge structure. *TRAC*, **11**(7):237–244, 1992.

[7] S. Behrens, H. Bönnemann, N. Matoussevitch, V. Kempter, W. Riehemann, A. Wiedenmann, S. Odenbach, S. Will, D. Eberbeck, R. Hergt, R. Müller, K. Landfester, A. Schmidt, D. Schüler, and R. Hempelmann. Synthesis and characterization. In S. Odenbach, editor, *Colloidal Magnetic Fluids: Basics, Development and Application of Ferrofluids*, pages 1–69. Springer, Berlin, Heidelberg, 2009.

[8] U. Bergmann, C. R. Horne, T. J. Collins, J. M. Workman, and S. P. Cramer. Chemical dependence of interatomic X-ray transition energies and intensities – a study of Mn $K\beta''$ and $K\beta_{2,5}$ spectra. *Chem. Phys. Lett.*, **302**(1–2):119–124, 1999.

[9] M. Birkholz. *Grazing Incidence Configurations, in "Thin Film Analysis by X-Ray Scattering"*. Wiley-VCH Verlag GmbH & Co. KGaA, Weinheim, Germany, 2006.

[10] H. Bönnemann, U. Endruschat, J. Hormes, G. Köhl, S. Kruse, H. Modrow, R. Mörtel, and K. S. Nagabhushana. Activation of Colloidal PtRu Fuel Cell Catalysts via a Thermal "Conditioning Proces". *Fuel Cells*, **4**(4):297–308, 2004.

[11] D. Bohm and D. Pines. A Collective Description of Electron Interactions: III. Coulomb Interactions in a Degenerate Electron Gas. *Phys. Rev.*, **92**(3):609–625, 1953.

[12] H. Bönnemann and G. Khelashvili. Efficient fuel cell catalysts emerging from organometallic chemistry. *Appl. Organometl. Chem.*, **24**:257–268, 2010.

[13] C. H. Booth and F. Bridges. Improved Self-Absorption Correction for Fluorescence Measurements of Extended X-Ray Absorption Fine-Structure. *Physica Scripta*, **T115**:202–204, 2005.

Bibliography

[14] W. L. Bragg. The diffraction of short electromagnetic waves by a crystal. *P. Camb. Philosophical Society*, 17:43–57, 1913.

[15] M. Cahay, J. P Leburton, D. J. Lockwood, S. Bandyopadhyay, and J. S. Harris, editors. *Quantum Confinement VI: Nanostructured Materials and Devices: Proceedings of the International Symposium*. Electrochemical Society, Pennington, N.J., 2001.

[16] J. Cai, P. Ruffieux, R. Jaafar, M. Bieri, T. Braun, S. Blankenburg, M. Muoth, A. P. Seitsonen, M. Saleh, X. Feng, K. Müllen, and R. Fasel. Atomically precise bottom-up fabrication of graphene nanoribbons. *Nature*, **466**:470–473, 2010.

[17] S. Chen, W. Sheng, N. Yabuuchi, P. J. Ferreira, L. F. Allard, and Y. Shao-Horn. Origin of Oxygen Reduction Reaction Activity on "Pt_3Co" Nanoparticles: Atomically Resolved Chemical Compositions and Structures. *J. Phys. Chem. C*, **113**(3):1109–1125, 2009.

[18] J. Clark and K. H. Jack. *Chem. and Ind.*, **51**:1004, 1951.

[19] F. de Groot, G. Vankó, and P. Glatzel. The 1s x-ray absorption pre-edge structures in transition metal oxides. *J. Phys.: Condens. Matter*, **21**(10):104207, 2009.

[20] F. M. F. de Groot. High-Resolution X-ray Emission and X-ray Absorption Spectroscopy. *Chem. Rev.*, **101**(6):1779–1808, 2001.

[21] F. M. F. de Groot. Multiplet effects in X-ray spectroscopy. *Coordin. Chem. Rev.*, **249**:31–63, 2005.

[22] D. P. Dinega and M. G. Bawendi. A Solution-Phase Chemical Approach to a New Crystal Structure of Cobalt. *Angew. Chem. Int. Ed.*, **38**(12):1788–1791, 1999.

[23] M. Gell-Mann and K. A. Brueckner. Correlation Energy of an Electron Gas at High Density. *Phys. Rev.*, **106**(2):364–368, 1957.

[24] O. Glatter and O. Kratky, editors. *Small Angle X-ray Scattering*. Academic Press Inc., London, 1982.

[25] P. Glatzel and U. Bergmann. High resolution 1s core hole X-ray spectroscopy in 3d transition metal complexes: electronic and structural information. *Coordin. Chem. Rev.*, **249**:65–95, 2005.

[26] P. Glatzel, L. Jacquamet, U. Bergmann, F. M. F. de Groot, and S. P. Cramer. Site-Selective EXAFS in Mixed-Valence Compounds Using High-Resolution Fluorescence Detection: A Study of Iron in Prussian Blue. *Inorg. Chem.*, **41**(12):3121–3127, 2002.

[27] P. Glatzel, M. Sikora, and M. Fernández-García. Resonant X-ray spectroscopy to study K absorption pre-edges in 3d transition metal compounds. *Eur. Phys. J. Special Topics*, **169**:207–214, 2009.

[28] P. Glatzel, M. Sikora, G. Smolentsev, and M. Fernández-García. Hard X-ray photon-in photon-out spectroscopy. *Catalysis Today*, **145**:294–299, 2009.

[29] P. Glatzel, G. Smolentsev, and G. Bunker. The electronic structure in 3d transition metal complexes: Can we measure oxidation states? *J. Phys: Conf. Ser.*, **190**:012046, 2009.

[30] M. M. Grush, G. Christou, K. Hämäläinen, and S. P. Cramer. Site-Selective XANES and EXAFS: A Demonstration with Manganese Mixtures and Mixed-Valence Complexes. *J. Am. Chem. Soc.*, **117**(21):5895–5896, 1995.

[31] R. Guirado-López, F. Aguilera-Granja, and J. M. Montejano-Carrizales. Electronic structure and stability of polycrystalline cobalt clusters. *Phys. Rev. B*, **65**:045420, 2002.

[32] S. J. Gurman. Interpretation of EXAFS Data. *J. Synchrotron Rad.*, **2**(1):56–63, 1995.

[33] HASYLAB. HASYLAB W1 beamline.

[34] HASYLAB. HASYLAB W1 spectrometer.

[35] HASYLAB. Synchrotron Radiation.

[36] B. W. Holland, J. B. Pendry, R. F. Pettifer, and J. Bordas. Atomic origin of structure in EXAFS experiments. *J. Phys. C: Solid State Phys.*, **11**(3):633–642, 1978.

[37] J. Hormes, H. Modrow, H. Bönnemann, and C. S. S. R. Kumar. The influence of various coatings on the electronic, magnetic, and geometric properties of cobalt nanoparticles (invited). *J. Appl. Phys.*, **97**:10R102, 2005.

[38] B. J. Hwang, Monalisa Ming-Yao Cheng Din-Goa Liu S. M. S. Kumar, C.-H. Chen, and Jyh-Fu Lee. Understanding of Adsorption and Catalytic Properties of Bimetallic Pt–Co Alloy Surfaces from First Principles: Insight from Disordered Alloy Surfaces. *Phys. Chem. C*, **114**(15):7141–7152, 2010.

[39] A. Ichimiya and P. I. Cohen. *Reflection high-energy electron diffraction*. Cambridge University Press, Cambridge, 2011.

[40] Y. Jean. *Molecular orbitals of transition metal complexes*. Oxford University Press, New York, 2005.

[41] H. H. Johann. Die Erzeugung lichtstarker Röntgenspektren mit Hilfe von Konkavkristallen. *Z. Phys.*, **69**(3–4):185–206, 1931.

[42] Y. Joly. X-ray absorption near-edge structure calculations beyond the muffin-tin approximation. *Phys. Rev. B*, **63**(12):125120, 2001.

[43] J. B. Jones and D. S. Urch. Metal-ligand bonding in some vanadium compounds: A study based on X-ray emission data. *J. Chem. Soc., Dalton Trans.*, page 1885, 1975.

[44] D. K. Kirui, D. A. Rey, and C. A Batt. Gold hybrid nanoparticles for targeted phototherapy and cancer imaging. *Nanotechnology*, 21(10):105105, 2010.

[45] J. R. Kitchin, J. K. Nørskov, M. A. Barteau, and J. G. Chen. Role of Strain and Ligand Effects in the Modification of the Electronic and Chemical Properties of Bimetallic Surfaces. *Phys. Rev. Lett.*, **93**(15):156801, 2004.

[46] A. Kotani and S. Shin. Resonant inelastic x-ray scattering spectra for electrons in solids. *Rev. Mod. Phys.*, **73**(1):203–246, 2001.

Bibliography

[47] H. A. Kramers and W. Heisenberg. über die streuung von strahlung durch atome. *Z. Phys.*, **31**(1):681–708, 1925.

[48] M. O. Krause and J. H. Oliver. Natural widths of atomic K and L levels, Kα X-ray lines and several KLL Auger lines. *J. Phys. Chem. Ref. Data*, **8**(2):329–339, 1979.

[49] T.-J. Kühn, W. Caliebe, N. Matoussevitch, Helmut Bönnemann, and J. Hormes. Site-selective X-ray absorption spectroscopy of cobalt nanoparticles. *Appl. Organometal. Chem.*, **25**(8):577–584, 2011.

[50] T.-J. Kühn, P. Glatzel, N. Matoussevitch, Helmut Bönnemann, and J. Hormes. Site-selective high-resolution x-ray absorption spectroscopy and high-resolution x-ray emission spectroscopy of cobalt nanoparticles. *submitted to J. Am. Chem. Soc.*, 2012.

[51] N. Lee, T. Petrenko, U. Bergmann, F. Neese, and S. DeBeer. Probing Valence Orbital Composition with Iron Kβ X-ray Emission Spectroscopy. *J. Am. Chem. Soc.*, **132**(28):9715–9727, 2010.

[52] M. Lemonnier, O. Collet, C. Depautex, J.-M. Esteva, and D. Raoux. High Vacuum two Crystal soft X-Ray Monochromator. *Nucl. Instrum. Methods*, **152**(1):109–111, 1978.

[53] P. Machek, W. Caliebe E. Welter, U. Brüggmann, G. Dräger, and M. Fröba. Johann Spectrometer for High Resolution X-Ray Spectroscopy. *AIP Conference Proceedings*, **879**:1755–1758, 2007.

[54] G. Meitzner. *In Situ XAS Characterization of Heterogeneous Catalysts, in "In-Situ Spectroscopy in Heterogeneous Catalysis"*. Wiley-VCH Verlag GmbH & Co. KGaA, Weinheim, Germany, 2004.

[55] A. Mihelič. XANES spectroscopy.

[56] H. Modrow, N. Palina, C. S. S. R. Kumar, E. E. Doomes, M. Aghasyan, V. Palshin, R. Tittsworth, J. C. Jiang, and J. Hormes. Characterization of Size Dependent Structural and Electronic Properties of CTAB-Stabilized Cobalt Nanoparticles by X-ray Absorption Spectroscopy. *Phys. Scr.*, **T115**:790–793, 2005.

[57] P. Müller-Buschmann. A basic introduction to grazing incidence small-angle x-ray scattering. In T. A. Ezquerra, MC. Garcia-Gutierrez, A. Nogales, and M. Gomez, editors, *Applications of Synchrotron Light to Scattering and Diffraction in Materials and Life Sciences*, chapter 3. Springer, Berlin, 2009.

[58] S. Nagakura. Study of Metallic Carbides by Electron Diffraction Part IV. Cobalt Carbides. *J. Phys. Soc. Jpn*, **16**:1213–1219, 1961.

[59] K. M. Nam, J. H. Shim, D.-W. Han, H. S. Kwon, Y.-M. Kang, Y. Li, H. Song, W. S. Seo, and J. T. Park. Syntheses and Characterization of Wurtzite CoO, Rocksalt CoO, and Spinel Co3O4 Nanocrystals: Their Interconversion and Tuning of Phase and Morphology. *Chem. Mater.*, **22**(15):4446–4454, 2010.

[60] M. Newville. *Fundamentals of XAFS*. Consortium for Advanced Radiation Sources, University of Chicago, Chicago, 2004.

[61] NIST. X-ray Transitions Energy Database.

[62] J. K. Nørskov, J. Rossmeisl, A. Logadottir, and L. Lindqvist. Origin of the Overpotential for Oxygen Reduction at a Fuel-Cell Cathode. *J. Phys. Chem. B.*, **108**(46):17886–17892, 2004.

[63] C. Nützenadel, A. Züttel, D. Chartouni, G. Schmid, , and L. Schlapbach. Critical size and surface effect of the hydrogen interaction of palladium clusters. *Eur. Phys. J. D*, **8**(2):245–250, 2000.

[64] J. B. Pendry. *Low energy electron diffraction: the theory and its application to determination of surface structure.* Academic Press, New York, 1974.

[65] S. J. Pennycook and P. D. Nellist, editors. *Scanning Transmission Electron Microscopy: Imaging and Analysis.* Springer, New York, 2011.

[66] V. Petkov, V. Buscaglia, M. T. Buscaglia, Z. Zhao, and Y. Ren. Structural coherence and ferroelectricity decay in submicron- and nano-sized perovskites. *Phys. Rev. B*, **78**(5):054107, 2008.

[67] C. J. Pollock and S. DeBeer. Valence-to-Core X-ray Emission Spectroscopy: A Sensitive Probe of the Nature of a Bound Ligand. *J. Am. Chem. Soc.*, **133**(14):5594–5601, 2011.

[68] B. Ravel. IFEFFIT: ATHENA documentation.

[69] B. Ravel and M. Newville. ATHENA, ARTEMIS, HEPHAESTUS: data analysis for X-ray absorption spectroscopy using IFEFFIT. *J. Synchrotron Rad.*, **12**:537–541, 2005.

[70] J. J. Rehr. FEFF Wiki.

[71] J. J. Rehr and R. C. Albers. Scattering-matrix formulation of curved-wave multiple-scattering theory: Application to x-ray absorption fine structure. *Phys. Rev. B*, **41**(12):8139–8149, 1990.

[72] J. J. Rehr, R. C. Albers, and S. I. Zabinsky. High-order multiple-scattering calculations of x-ray-absorption fine structure. *Phys. Rev. Lett.*, **69**(23):3397–3400, 1992.

[73] J. J. Rehr, C. H. Booth, F. Bridges, and S. I. Zabinsky. X-ray-absorption fine structure in embedded atoms. *Phys. Rev. B*, **49**(17):12347–12350, 1994.

[74] J. J. Rehr and R. C. Albers. Theoretical approaches to x-ray absorption fine structure. *Rev. Mod. Phys.*, **72**(3):621–654, 2000.

[75] J. J. Rehr, J. J. Kas, M. P. Prange, A. P. Sorini, Y. Takimoto, and F. Vila. Ab initio theory and calculations of X-ray spectra. *C. R. Physique*, **10**(6):548–559, 2008.

[76] M. Respaud, J. M. Broto, H. Rakoto, A. R. Fert, L. Thomas, B. Barbara, M. Verelst, E. Snoeck, P. Lecante, A. Mosset, J. Osuna, T. Ould Ely, C. Amiens, and B. Chaudret. Surface effects on the magnetic properties of ultrafine cobalt particles. *Phys. Rev. B*, **57**(5):2925–2935, 1998.

[77] S. Rudenkiy, M. Frerichs, F. Voigts, W. Maus-Friedrichs, V. Kempter, R. Brinkmann, N. Matoussevitch, W. Brijoux, H. Bönnemann, N. Palina, and H. Modrow. Study of the structure and stability of cobalt nanoparticles for ferrofluidic applications. *Appl. Organometal. Chem.*, **18**:553–560, 2004.

Bibliography

[78] V. A. Safonov, L. N. Vykhodtseva, Y. M. Polukarov, O. V. Safonova, G. Smolentsev, M. Sikora, S. G. Eeckhout, and P. Glatzel. Valence-to-Core X-ray Emission Spectroscopy Identification of Carbide Compounds in Nanocrystalline Cr Coatings Deposited from Cr(III) Electrolytes Containing Organic Substances. *J. Phys Chem B*, **110**(46):23192–23196, 2006.

[79] J. J. Sakurai. *Modern Quantum Mechanics*. Addison Wesley Publishing Company, New York, 1994.

[80] D. E. Sayers, E. A. Stern, and F. W. Lytle. New Technique for Investigating Noncrystalline Structures: Fourier Analysis of the Extended X-Ray–Absorption Fine Structure. *Phys. Rev. Lett.*, **27**(18):1204–1207, 1971.

[81] H. Schulenburg, E. Müller, G. Khelashvili, T. Roser, H. Bönnemann, A. Wokaun, and G. G. Scherer. Heat-Treated PtCo$_3$ Nanoparticles as Oxygen Reduction Catalysts. *J. Phys. Chem. C*, **113**(10):4069–4077, 2009.

[82] W. S. Seo, J. H. Shim, S. J. Oh, E. K. Lee, N. H. Hur, and J. T. Park. Phase- and Size-Controlled Synthesis of Hexagonal and Cubic CoO Nanocrystals. *J. Am. Chem. Soc.*, **127**:6188–6189, 2005.

[83] W. L. Smith and A. D. Hobson. The structure of cobalt oxide, Co$_3$O$_4$. *Acta Cryst. B*, **29**:362–363, 1973.

[84] G. Smolentsev, G. Guilera, M. Tromp, S. Pascarelli, and A. V. Soldatov. Local structure of reaction intermediates probed by time-resolved x-ray absorption near edge structure spectroscopy. *J. Chem. Phys.*, **130**(17):174508, 2009.

[85] G. Smolentsev, A. V. Soldatov, J. Messinger, K. Merz, T. Weyhermüller, U. Bergmann, Y. Pushkar, J. Yano, V. K. Yachandra, and P. Glatzel. X-ray Emission Spectroscopy to Study Ligand Valence Orbitals in Mn Coordination Complexes. *J. Am. Chem. Soc.*, **131**(36):13161–13167, 2009.

[86] Y. Song, H. Modrow, L. L. Henry, C. K. Saw, E. E. Doomes, V. Palshin, J. Hormes, and C. S. S. R. Kumar. Microfluidic Synthesis of Cobalt Nanoparticles. *Chem. Mater.*, **18**(12):2817–2827, 2006.

[87] S. Sun and C. B. Murray. Synthesis of monodisperse cobalt nanocrystals and their assembly into magnetic superlattices (invited). *J. Appl. Phys.*, **85**(8):4325–4330, 1999.

[88] J. C. Swarbrick, Y. Kvashnin, K. Schulte, K. Seenivasan, C. Lamberti, and P. Glatzel. Ligand Identification in Titanium Complexes Using X-ray Valence-to-Core Emission Spectroscopy. *Inorg. Chem.*, **49**(18):8323–8332, 2010.

[89] J. C. Swarbrick, U. Skyllberg, T. Karlsson, and P. Glatzel. High Energy Resolution X-ray Absorption Spectroscopy of Environmentally Relevant Lead(II) Compounds. *Inorg. chem.*, **48**(22):10748–10756, 2009.

[90] B. Thiesen and A. Jordan. Clinical applications of magnetic nanoparticles for hyperthermia. *Int. J. Hyperthermia*, 24(6):467–474, 2008.

[91] J. Tollefson. Hydrogen vehicles: Fuel of the future? *Nature*, **464**:1262–1264, 2010.

[92] A. Ulman. Formation and Structure of Self-Assembled Monolayers. *Chem. Rev.*, **96**(4):1533–1554, 1996.

[93] UnderstandingNano.com. Nanotechnology Applications.

[94] D. S. Urch. *Electron spectroscopy: Theory, Techniques, and Applications.* Academic Press, New York, 1979.

[95] U. von Barth and G. Grossmann. Dynamical effects in x-ray spectra and the final-state rule. *Phys. Rev. B*, **25**(8):5150–5179, 1982.

[96] E. Welter, P. Machek, G. Dräger, U. Brüggmann, and M. Fröba. A new X-ray spectrometer with large focusing crystal analyzer. *J. Synch. Rad.*, **12**(4):448–454, 2005.

[97] F. Wen and Helmut Bönnemann. A facile one-pot synthesis of [(COD)Pt(CH$_3$)$_2$]. *Appl. Organomet. Chem.*, 19(1):94–97, 2005.

[98] Wikipedia. Beer-Lambert law.

[99] Wikipedia. Bragg's law.

[100] Wikipedia. X-ray absorption.

[101] Wikipedia. Fuel cell.

[102] Wikipedia. Nanotechnology applications.

[103] Wikipedia. Nanoparticle.

[104] Wikipedia. Nanotechnology.

[105] Wikipedia. Proton exchange membrane fuel cell.

[106] Wikipedia. Singular Value Decomposition.

[107] Wikipedia. Synchrotron.

[108] K. Wille. *The physics of particle accelerators: an introduction.* Oxford University Press, New York, 1996.

[109] D. B. Williams and C. Barry Carter, editors. *Transmission Electron Microscopy: A Textbook for Materials Science.* Springer, New York, 2009.

[110] Z. Y. Wu, D. C. Xian, T. D. Hu, Y. N. Xie, Y. Tao, C. R. Natoli, E. Paris, and A. Marcelli. Quadrupolar transitions and medium-range-order effects in metal K-edge x-ray absorption spectra of 3d transition-metal compounds. *Phys. Rev. B*, **70**:033104, 2004.

[111] Y. Xu, A. V. Ruban, and M. Mavrikakis. Adsorption and Dissociation of O_2 on Pt-Co and Pt-Fe Alloys. *J. Am. Chem Soc.*, **126**(14):4717–4725, 2004.

[112] J. S. Yin and Z. L. Wang. Ordered Self-Assembling of Tetrahedral Oxide Nanocrystals. *Phys. Rev. Lett.*, **79**(13):2570–2573, 1997.

Bibliography

[113] F. Zasada, W. Piskorz, P. Stelmachowski, A. Kotarba, J.-F. Paul, T. Plociski, K. J. Kurzydlowski, and Z. Sojka. Periodic DFT and HR-STEM Studies of Surface Structure and Morphology of Cobalt Spinel Nanocrystals. Retrieving 3D Shapes from 2D Images. *J. Phys. Chem. C*, **115**(14):6423–6432, 2011.

[114] P. Zhang and T. K. Sham. X-Ray Studies of the Structure and Electronic Behavior of Alkanethiolate-Capped Gold Nanoparticles: The Interplay of Size and Surface Effects. *Phys. Rev. Lett.*, **90**(24):245502, 2003.

i want morebooks!

Buy your books fast and straightforward online - at one of world's fastest growing online book stores! Environmentally sound due to Print-on-Demand technologies.

Buy your books online at
www.get-morebooks.com

Kaufen Sie Ihre Bücher schnell und unkompliziert online – auf einer der am schnellsten wachsenden Buchhandelsplattformen weltweit! Dank Print-On-Demand umwelt- und ressourcenschonend produziert.

Bücher schneller online kaufen
www.morebooks.de

 VDM Verlagsservicegesellschaft mbH
Heinrich-Böcking-Str. 6-8   Telefon: +49 681 3720 174   info@vdm-vsg.de
D - 66121 Saarbrücken   Telefax: +49 681 3720 1749   www.vdm-vsg.de

Printed by Books on Demand GmbH, Norderstedt / Germany